INTERIBERICA, S.A. DE EDICIONES

LIFE IN THE FUTURE

Prospects for Man and Nature

Doubleday and Company Inc.,
Garden City, New York, 1976

Also published in parts as Life in the Future
and Ecologists at Work

LIFE
IN THE FUTURE

by Malcolm Ross-Macdonald

An 8-year-old child's view of the future, heavily dominated by space and communications hardware.

Series Coordinator	Geoffrey Rogers
Art Director	Frank Fry
Design Consultant	Guenther Radtke
Editorial Consultant	Donald Berwick
Series Consultant	Malcolm Ross-Macdonald
Editors	Damian Grint
	Maureen Cartwright
Research	Enid Moore
	Peggy Jones
	Barbara Fraser
Art Assistant	Michael Turner

ISBN: 84-382-0022-2 Dep. Legal: S.S. 809-1975
Printed in Spain by TONSA - San Sebastián

Contents

Ecologists at Work

A supplementary account of the experimental work of ecologists, prepared by Michael Hassell and Stuart McNeill

Editorial Advisers

DAVID ATTENBOROUGH. Naturalist and Broadcaster

MICHAEL BOORER, B.SC. Author, Lecturer and Broadcaster

Foreword by David Attenborough

*P*rophets of disaster have never been in short supply. Back in Old Testament times, holy men were predicting the imminence of hell-fire and destruction. Two thousand years later, others were still doing much the same thing and, what is more, going so far as to specify the precise calendar date of doomsday. Yet, in spite of such fearful predictions over so many centuries, the world continues to spin. The fact that it does so, of course, is no guarantee that catastrophe will not ultimately overwhelm us, and unhappily there are reasons why the forecasts of disaster made today must be taken more seriously than ever before. The ancients believed that the world would be brought to an end by divine intervention. None foretold that man would have the power to bring about such a thing himself. But that, now, is the case.

Today, we have poisons so lethal that we can not only exterminate creatures we consider a nuisance, but dispense death, inadvertently, far beyond our intended targets. We have built machinery so powerful that with it a few men in a few days can annihilate a forest. We have devised mechanical methods to plunder sources of energy that have taken millions of years to accumulate, and invented devices that can consume it all in decades. And we have in our hands atomic weapons that can devastate continents. Little wonder that forecasts of disaster have recently become so loud, so frightening, and so plausible. Malcolm Ross-Macdonald, in the first chapters of this book, takes a sober and analytical look at such forecasts, at the facts on which they are based and those that they ignore.

The discovery of such facts is the second reason why the predictions made today differ so importantly from the warnings of earlier times. Contemporary prophets extrapolate, not from private revelations but from verifiable observations. Michael Hassell and Stuart McNeill, in the second half of the book, detail some of the many methods that ecologists have devised to collect their data. Their conclusions not only enable us to understand the complex interrelationships that bind us and all living things into one great web, but also make it possible for us to predict, with a greater chance of accuracy than ever before, the consequences of our activities. If we take such knowledge to heart and allow the insights of ecology to temper our thoughtlessness and our greed, then we may yet manage to confound the worst nightmares of the prophets of doom.

David Attenborough.

The Fascination of the Future

There is no doubt that the future is, and always has been, one of the most seductive subjects to which the human mind can turn. Anyone who has ever whiled away a dull or sleepless hour in the pastime of planning—whether for next summer's flower garden, a new home, a fortune-spinning business, or a vacation—will know just how engaging the future becomes when reality can be completely ignored or temporarily tamed into compliance with fantasy.

Fortunately, maturity brings a healthy skepticism to temper that kind of euphoria. Experience teaches that tomorrow's gardens, homes, business deals, and vacations turn out to be pretty much like yesterday's. We can make believe it will be otherwise without too much harm; but if we *truly* believe it, we are behaving more like compulsive gamblers than like shrewd managers of affairs. Then, too, the nature of an unbridled daydream, the fact that it occupies time out of our more sober and useful workday lives, helps to devaluate it in advance.

But what of the futurologists—those who make a profession of forecasting the future? Around the world there must by now be several thousand practitioners of this new science, men and women who devote their working lives to devising and refining techniques for predicting the broad future. And for every one of these there are a further 1000, if not 10,000, whose concern is some particular aspect of that broad future. Their field may be global, such as the future of the energy industries, or it may be literally parochial, such as the traffic pattern around the village green five years from now. The futurologists are people with a full- or part-time professional concern with the future; someone pays them to make predictions and suggest courses of action on the basis of their expectations.

The predictions they make are not of the daydream kind (though "daytime nightmare" might

For most people, experts as well as laymen, the idea of the future stretches forward only half a century or so—which is more or less the lifetime of a typical family unit. Beyond that span few experts are willing to make serious predictions and few laymen feel any great interest or personal concern.

Left: in each of us there is something that responds to, and may even thrill to, disaster or tragedy. In the humor of Charles Addams that element is merely taken to its comic extreme, but it still touches some cord within us.

Right: much of science fiction appeals to our sense of fear and menace concerning the future, which is the ultimate unknown and unknowable. But its imagery has always been firmly rooted in the present, as in this 1930s vision of a "menace from space" shown here.

Below: one product of recent fears of the chance of a thermonuclear world war was the film Dr. Strangelove, or How I Learned to Stop Worrying and Love The Bomb—*a tense moment from which is shown here, with the US President and the Soviet Ambassador on the hot line.*

Drawing by Chas. Addams; © 1946, 1974 The New Yorker Magazine Inc.

TALES of WONDER

N.º 4

1/-

THE MENACE FROM SPACE

by JOHN EDWARDS

aptly characterize some of their findings), nor are they much concerned with the utterly un- guessable long-term future—a century or more ahead, say. As for the *very* long-term future, we actually know a great deal about it, and there is little point in concerning ourselves with it. Here, for instance, are some of its certainties:

The sun will explode spectacularly as either a nova or a supernova, or else it will swell and cool slightly, becoming a red giant. Either type of event will destroy the earth as we know it.

Earth and moon will collide.

The present arrangement of continents and oceans will change beyond recognition.

Most, if not all, of the existing species will eventually become extinct, supplanted by the new forms that will evolve from the present ones.

Predictions of such far-off and inescapable happenings make little impression upon us, and can in no way condition our daily lives. Nobody is likely to change his ways because earth and moon will one day collide; few would pay more for hilltop land because it will be spared from flooding when Antarctica melts. But the short-term future, upon which the futurologists focus most of their attention, is quite another thing. If we do not manage to survive the short term— the next 10 to 25 years—there can be no long-term worries anyhow.

Those words "if we do not survive" are pointers to something entirely new in the long history of man's speculations about the future. True, the

Traditional "futurologists" made faith and magic—rather than reason and science— the basis of their systems, which included casting "wise" stones (left), laying out cards whose face markings and sequence are invested with significance (center), and gazing into crystal balls (right). In times of great uncertainty such astrological arts enjoy a widespread and spontaneous revival.

human mind seems always to have been haunted by the notion of an apocalypse—a time of universal destruction in fire or flood. Such visions form the cornerstone of many faiths. In times of upheaval and social turmoil, people seem to become receptive to the message that they are damned, that the end is nigh. So we who live at such a time should take our skepticism, as well as our courage, in both hands as we face our daily ration of doomsaying. Yet, undeniably, the doomsayers were never before able to put together such a convincing case as they now can. They need not feed us vague quotes from purposely ambiguous oracles, nor need they rely on the approach of a centennial or millennial year or on the births of two-headed calves and other wonders. Today's prophets can point to actual systems that can annihilate all human life and most other life as well. And at various levels below that ultimate horror we can discern threats of diminishing magnitude, the least of which is, nonetheless, more frightening and global than anything in prior history.

Some of the possibilities within other possibilities do not involve total death or destruction. In a way, though, these can be even more alarming than the others. Death, after all, is a grisly but final solution to every problem. But survival in a world whose economy and institutions have collapsed could bring into being an unimaginable universal savagery and barbarism.

The paragraphs you have just read describe,

in terms kept deliberately vague, a world civilization in crisis. I have been talking about threats without actually naming one of them. Yet there can be few modern readers whose minds do not call up a number of specific terms—ready, as it were, to match the expected printed word: nuclear holocaust, economic collapse, biological genocide, starvation, population explosion, pollution, depleted resources. . . . If such terms have not been in your mind as you read, you have been amazingly untouched by the thinking that has crammed the media for at least a decade.

Our modern vision of the apocalypse grows out of these specific fears of what *is* happening or really *could* happen. This gives it quite a different character from previous visions, which had their authority in allegedly divine messages, sometimes communicated through readings of the stars or of chickens' entrails. In other words, there is no element of the supernatural in the fears that now concern us; they are based on observable, predictable processes. And this book is about that kind of observation and that kind of prediction. It is not a collection of oracles.

We want to know as much as possible about the future now, so that we can be ready for it when it comes. We do not want to build highways, or colleges, or homes, or shops where there will be no demand for them; on the other hand, we do not want our existing facilities to burst at the seams because we failed to realize they would need expanding. In the past, people have not been notably foresighted. When Europe and America were fast industrializing during the 19th century, each town rushed up hundreds of acres of substandard housing without giving a thought even to such elementary things as drainage. In London in the 1840s they actually had to pass laws forbidding people to discharge their new-fangled water closets into the city's sewers, which had been designed to take only the runoff from kitchen sinks and an infrequent bathtub. Cesspools were still being emptied by "night men," who put bucketful after bucketful into carts for transportation out to farms. The poor could not afford this service and their night soil as well as other domestic garbage was heaped high on

Around the turn of the last century many studies showed that most social evils were related to poverty, and that piecemeal reforms were unlikely to succeed. Movements arose then to change the whole basis of society: some by revolution, others— such as the Salvation Army—peacefully and constitutionally.

vacant lots, often high enough to tower over the crowded nearby houses.

It is past experience of this kind that has helped us to realize the value of being able to predict the near future, and so to prepare for it. (Incidentally, the example shows, too, that the problem of population in the cities is by no means new.) Fortunately, we are better equipped for the business of prediction than were our 19th-century forebears. When we look back at those days, we can see a steady increase in most of the positive social qualities: national wealth, personal income, standard of living, schools, life expectancy, literacy, and so on. But it did not appear like that to people alive then. In the thick of the wood, although you can see individual trees clearly, you may get no inkling of the full shape of the forest; unless you take a measuring tape and surveyor's kit, you may still have the wrong idea after several walks within and around the wood. In the 19th century—and, of course, in earlier times—there was a shortage of good tools for measuring social qualities quickly enough for the results to be of use in planning for the future. We now have more such tools than we have ever had before.

We are also the inheritors of a long tradition of social improvement. We take social improvement for granted, and it seems natural to us to plan ahead to meet it. But those who lived much nearer to the start of the tradition saw only an endless succession of booms and slumps, because they knew only the crudest techniques for relating the supply of money to all other forms of economic activity. Would this year's boom last? Certainly not. How long would the coming slump endure? Nobody could say. How could one even think of planning when the very basis of the plans shifted wildly from year to year? Instead, men took economic and social life as they took the weather: it was just one of those things that they could do little about.

But our great-grandparents learned fast. Before the end of the century, men such as Charles Cooley and Lester Ward in America, Charles Booth and B. S. Rowntree in England, A. J. L. Quételet in Belgium, and many others had laid the foundations of the scientific study of society as we know such study today. Charles Booth, for instance, overturned the central belief of most 19th-century thinkers that poverty resulted primarily from vice, drunkenness, or laziness, and that poverty was the proper and just reward for moral lapses. He showed that its chief causes were unemployment, death of bread-winners, trade depression, old age, and accident. Naturally people had always known that such things led to *some* economic suffering. But it shattered their complacency to learn that these were the chief causes. Booth's story shows how quickly things can be made to change, too; in little over a decade after his findings came out, Britain initiated systems of old-age pensions, government employment offices, unemployment benefits, and similar reforms.

Pioneer studies such as Charles Booth's brought about two fundamental changes. First, they turned what had previously been mere hunches or suspicions or insights about society into hard facts. They brought the possibility of *measurement* into the very heart of social study. Until then, social statistics had been a handy but fairly minor tool. Secondly, and even more important from our present point of view, they stopped men from thinking that social forces were, like the weather, something to put up with; more and more, people came to realize that they could control and even at times create social forces. They began to see society as something they could deliberately mold.

The early days of social science were heady, idealistic days of great fervor—too much fervor, perhaps. One possible definition of mankind is "the animal that always goes too far." You name it, we have overdone it—somewhere, sometime. Two world wars and the swallowing-up of idealism in one dictatorship after another have tempered both our enthusiasm for molding and creating social forces and our faith in science-directed social progress. Still, the grim 20th-century years have left untouched the basic core of our belief that social forces can at least be measured and in some degree channeled. Hand in hand with such a belief goes the notion that the same forces can also be predicted for a useful distance into the future.

These ideas now seem so obvious that it comes as something of a shock to realize that for most of mankind's history they simply did not occur to anybody. Even during our recent history—the last 10,000 years, say, since we became organized around town-centered civilizations—people have believed that societies go in cycles rather than evolve. The twin ideas of evolution and progress are only a century or so old. Let us look a little more deeply at how these ideas first emerged.

In the centuries before modern industry was born, the major theory of social change was one that had been devised by the Greek philosopher Aristotle (384–322 B.C.). After studying the constitutions of 158 states, past and current, in and around the Mediterranean, Aristotle concluded that states pass through an endless cycle, each stage of which contains the seed of its own destruction and thus prepares the way for the next stage. His cycle, as interpreted in modern phraseology, looks like this:

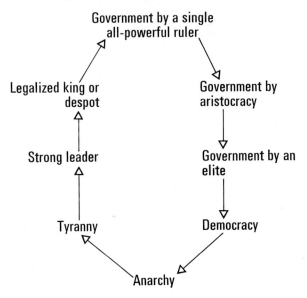

And the basic intellectual attitudes that accompany each phase are:

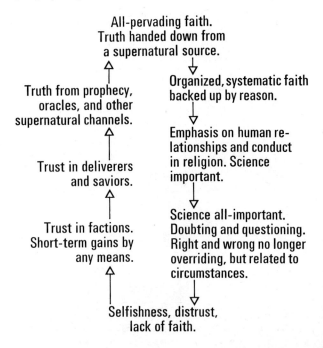

16

This is not the place for a discussion of whether or not Aristotle's theory is "true." It was obviously applicable to the majority of states he examined, or he would never have been led to these conclusions. Again, some parts of the cycle are certainly true of recent European history, other parts of recent American history, yet others of the communist world. But the kind of truth they involve is like that of the long-range weather forecast: given a certain pattern of weather, the next *most likely* development is such-and-such— or, given a recent pattern of social change, the next most likely stage is . . . whatever it may be.

Aristotle is important today not because his work serves as an aid to prediction, but because he really founded the systematic study of human society. Through his enormous influence, the thought has become firmly established in the collective Western consciousness that societies change in an ordered manner, and that there are underlying rules to govern the process. Once you accept such a belief, it follows that if you are observant, or assiduous, or perhaps just lucky, you can find the rules and will be able to devise a powerful new forecasting technology, as a result of which you and your fellow men will have gained some degree of control over that vast area of life hitherto reserved for palmists, astrologers, oracles, and readers of chickens' entrails.

The evolutionary theories of the 19th century, of which Darwin's was the most important, had a powerful influence on sociology. In fact, one could say that modern sociology started when evolutionary ideas cut Aristotle's cycle open and laid it out in the form of a string or chain of stages—an evolutionary chain. Societies, it now seemed, were not condemned to tread an endless circle. A spiral, yes, but not a circle. The possibility of progress and of evolution toward higher, better, more nearly perfect forms was now built into the modern theory of social change. And the times themselves lent color to the new ideas of progress. In 1750 six children out of 10 died before the age of five; a century later, thanks to improvements in sanitation—and wages—population everywhere was rising fast; in Britain, then leader among industrializing nations, it had doubled. It was possible for everyone to see a whole range of improvements in a single lifetime. People could also point to a new force—industry— that was creating wealth faster than anyone had believed possible.

Most important of all, these developments led to an intellectual climate in which expectations were altered. People *expected* "futurity" to be not only different but also much better. The English historian Thomas Babington Macaulay summed it up when he wrote in 1850: "Science has lengthened life; it has mitigated pain; it has extinguished diseases; it has increased the fertility of the soil; it has given new securities to the mariner; it has furnished new arms to the warrior; it has spanned great rivers and estuaries with bridges of form unknown to our fathers . . . it is a philosophy which never rests, which has never attained, which is never perfect. Its law is progress. A point which yesterday was invisible is its goal today, and will be its starting point tomorrow."

A century later, optimism of that kind was still very common, though by 1950 the voices of doubt and caution were beginning to rise to their present crescendo. But the doubters and pessimists, too, have a venerable history—all the way back to 1798, when an English clergyman, Thomas Robert Malthus, looked at the already conspicuous increase in birthrate, did some calculations using arithmetic and geometric progressions (which describe accelerating increases), and concluded that such goals as universal prosperity, happiness, and abundance are mirages that recede as fast as we approach them. He predicted "famine, distress, havoc, and dismay," and his version of bleak futurity became almost as commonplace in people's thinking as the other, optimistic view.

Human populations, Malthus concluded, would increase to the point where large numbers would starve, and surpluses of people would have to be removed by periodic wars, plagues, and famines. In the end, he said, human populations are limited by the yields from the soil: "When acre has been added to acre till all the fertile land is occupied, the yearly increase of food must depend on the melioration of the land already in possession. This is a fund, which, from the nature of all soils, instead of increasing, must gradually diminishing."

Malthus's great contribution in his own day was to show people that they could use mathematics as a means of predicting some future state of society. This was something quite new; but it soon became standard practice even in rather humdrum applications. By the middle of the 19th century, nobody would have dreamed of building even the most modest branch railway line without making a mathematical projection

Government by a man-god

All power, all truth, all wisdom, all law are in the hands of one man, considered divine or infallible. Skepticism may be punished by death.

Government by a legalized despot

Reason, science, and law take a back seat to revelation, prophecy, and royal decree. Power is completely centralized.

Aristotle's theory of social change

Government by a "messiah"

Tyranny gets institutionalized. Strong leaders rule through factions or parties whose actions the law (if it survives) merely rubber-stamps.

Tyranny

Society divides into factions unwilling to cooperate, having no beliefs or aims in common. Opportunistic strong men take advantage of the breakdown in law.

Government by aristocracy

Religious faith is organized and systematic, sustained by appeals to reason. Impartial law has some importance to government.

Government by an elite

Reason becomes more important than faith. Impartial law gains in importance—and so does science. But skepticism is barely welcome.

Democracy

Impartial law is the basis of government; not even the highest may flout it. Science is all-important. Skepticism is encouraged.

Anarchy

Skepticism boils over into cynicism and alienation. Despair replaces faith. Lynch law and violence are widely resorted to.

19

CHANTIER DE
DEMOLITIONS

of its freight and passenger potential first.

Nevertheless, wherever industry came, it seemed to completely upset Malthus's predictions, for populations grew and flourished. And the undoubted benefits of industrialism influenced most of the early commentators on society — men such as Saint-Simon, Comte, and de Tocqueville in France; Spencer and Marx in England; and Weber and Durkheim in Germany. Their work spans the century and naturally shows huge differences (Marx, for instance, was a founder of communism, but de Tocqueville became a bitter opponent of republican extremists and socialists). Even so, all of them saw their main role as that of interpreter; they aimed to explain the evolution of industrialism. They all believed that the many and varied societies of the world were converging toward industrialization and that Europe and America were merely the first through the hoop. As Marx said: "The country that is more developed industrially only shows, to the less developed, the image of its own future." It was his belief that industrial society was, either in fact or potentially, the best and highest type of society. Any talk of a long-ago Golden Age was relegated to the realm of myth.

The 19th century ended, then, with a number of writers expounding various theories of social change, but all the theories were based on a belief in social evolution. Some were optimistic, some pessimistic, but each offered a general explanation of how societies evolve.

In any science a good general theory is the handiest tool possible. Not only does it link a wider number of seemingly random facts and deductions into one coherent framework, but it also acts as a powerful aid to prediction. For instance, if you wanted to find out whether there is a planet beyond the known series, you could divide the night sky among several hundred astronomers, tell them to keep their eyes peeled, and hope for the best. But it would be more fruitful to turn to gravitational theory, which predicts that if there was a further planet out there, it would cause detectable movements in the orbit of the present outermost known planets. Indeed, that is exactly how the French astronomer Leverrier predicted in 1846 that a planet would be discovered beyond Uranus, and how the German astronomer Galle discovered it (we now call it Neptune) in the following year.

If sociologists had gone on developing their new science along the lines established in the 19th century, it is barely possible that we might by now have a meaningful theory of social change that would let us work out some general predictions. But they did not. In the early years of this present century, sociology took a sharp turn away from such general theorizing, and especially away from any notion of evolution as applied to social change.

There were good historical reasons for this switch. In the first place, industrial society was itself changing so fast that merely the study of it had become a full-time job. Facts piled up too fast and changed too swiftly to allow any breathing space for general theory building. And, secondly, the notion of social evolution had grown out of the narrow study of European society during the years of the Industrial Revolution and immediately before it — a very small part of the earth, and a very narrow window in time. What had been learned was of little help in the study of supposedly "primitive" societies past and present. In fact, it proved to be nothing but a hindrance.

All the talk about evolution from "primitive" to "advanced" societies was clouding our vision and leading to a lot of very biased reporting, especially of the most "primitive" societies. When 20th-century sociologists saw that this was so, they reacted strongly. Indeed, the pendulum swung so far that "evolution" became a dirty word; the nastiest name one sociologist could call another was "evolutionist." And, with few exceptions, that attitude still prevails. The study of social change has turned into a piecemeal affair, limited to one aspect, one suburb, one industry, one region. Generalizations about progress, which might lead down into the pit of evolutionism, are taboo.

This embargo on evolution-tainted theories brought many benefits, not least of which is a much better understanding of the marvelous organization and skill that go into what were

Visions of a bleak future ruined by "progress" are not especially modern. This scene—taken from a French book of the 1880s predicting life in the century to come—shows the old ideals of art and beauty being dynamited into oblivion. Presiding over the scene is the Future, a specter composed of chemical retort, cogwheels, an electric lamp, and other symbols of technology.

once dismissed arrogantly as "primitive" societies. But it has left us without any workable theory of social change.

This proved an enormous drawback when, after World War II, dozens of newly independent nations wanted to start planning for their own futures. For practical purposes, sociologists had to fall back on the nearest handy theories of change, which happened to be the apparently discredited theories of the evolutionists. Having nothing better of their own, the postwar planners blew the dust off them and began to apply them. Throughout the industrialized world, whether capitalist (called "market economies") or communist (called "centrally planned economies"), there was general agreement that the newly independent countries should evolve toward industrial states. At various times, they have

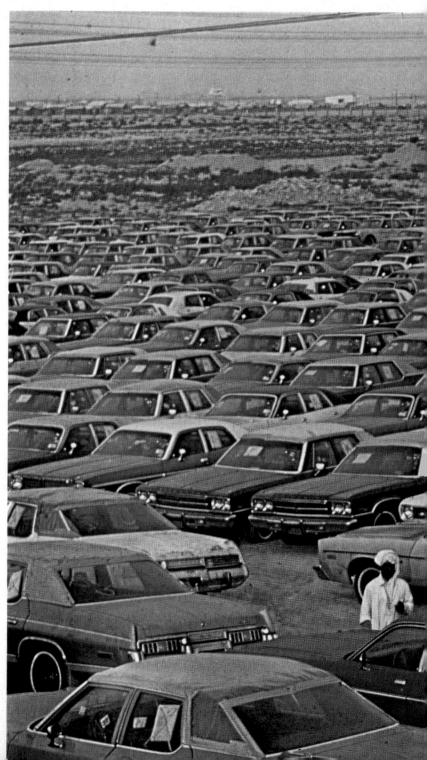

Less than a quarter-century ago the states of the Middle East were classed among the "backward" countries. Now, thanks to the dependence of the West on oil, they are among the richest nations on earth, though in terms of technology they are still extremely underdeveloped. For manufactures they depend almost entirely on imports, as this vast tract of imported cars in Kuwait suggests. By the turn of the century, when their oil runs out, they will have to create a diverse industrial economy if they want to maintain the standards their present wealth permits them to enjoy.

been called the "backward," "underdeveloped," or "developing" countries; one futurologist, Paul Ehrlich, has wryly suggested "never-to-be-developed" as the most suitable name. Today, though, they generally go by the neutral title "Third World," which has the merit of avoiding any suggestion of a commitment to the evolutionary thinking behind such words as "backward" or "developing."

Our disappointment over modern sociology's failure to come up with any general theory of social change—as well as our disenchantment with the somewhat cobwebby notions of 19th-century evolutionists, hastily updated in the 1950s—has left us rather adrift. Those of us who would like to guess the shape of tomorrow have little more than some rule-of-thumb techniques to guide us in our attempt.

How We Guess the Future

Let us imagine that one of your kindly ancestors invested the Roman equivalent of a dollar at a miserly 1 per cent interest compounded annually, beginning on the day in 44 B.C. when Julius Caesar was murdered in Rome. This is the way in which your imaginary inheritance has increased during the intervening centuries:

By the year A.D. 1 it was worth a whole 55 cents more. In something less than another seven centuries it crept up to a total of $1059. Then it started to pick up speed. By 1066, the year of the Norman Conquest of England, it was worth $62,670. When Columbus first saw America in 1492, it had swollen to $4,343,884. Less than

The drawings at the foot of these pages illustrate how one dollar, invested at 1 per cent per annum in 44 B.C., would have grown over the intervening years, and on into the future, to a sum that would make any taxman smile and leap into action.

44 B.C. $1.00

A.D.1 $1.55

1492 $4,343,884

1783 $73,316,331

three centuries later, when America won her independence, you had a respectable $73,316,331 fortune. The year of Britain's Great Exhibition, 1851, brought the figure to $154,638,206. By 1970 your ancestor's forethought had netted you a cool $505,328,006. In the first 44 years of the $1 investment, remember, it added a mere half-dollar to its value. In the 44 years from 1970 to 2014 it will add another $277,000,000. Give it one more millennium and it will have soared to over 16 trillion dollars.

Plot all these values on a graph and you get a curve whose shape has become familiar to 20th-century eyes: the exponential curve. If your ancestor had driven a harder bargain and secured a 2½ per cent interest rate, it would not have altered the general shape of the curve, but it would have sent it soaring quicker: in fact, you would have been a trillionaire by the 12th century instead of having to wait until after the year 2700. But the general nature of the curve would, nevertheless, be unaltered. All exponential curves have basically the same sort of shape—an early phase where the slope is low and growth is very sluggish, an intermediate phase during which there is an acceleration of lift-off, and then finally a runaway phase when the quantities multiply with dizzying speed.

A.D. 655 $1059

1066 $62,670

1851 $154,638,206

1970 $505,328,006

Curves of this kind describe the unhindered growth of virtually anything, any value, any quantity. Random examples would be: bacteria until they ran out of food; passenger miles flown by air until this present decade (and making allowance for two world wars); the increasing number of scientists in our century; the number of horsepower applied in industry from the 1780s until the 1960s; computer capacity; tourism; automobile sales until a few years ago; current world population trends . . . the potential list is vast. And such curves also describe similar growth situations in history—for example, the mileage of telegraph cables laid from the 1840s onward; the registered world tonnage of shipping from the 17th century to the 20th century; the growing wealth of the developed countries through the 19th century almost up to the present day . . . and population again.

Because the exponential curve applies so widely, it is naturally a very powerful aid to predicting the future of a large number of different activities. The history of scientific prediction is filled with examples of forecasters who have grasped the principle of exponential growth, have realized that it explains startling recent increases in some activity or another, and have then used the curve to predict the forward course of that activity. Often they have realized with a shock that the accelerating curve threatened to shoot out through the ceiling—and the shock has sometimes been great enough to turn them into militant crusaders bearing warnings of imminent collapse to those who have not yet grasped the runaway nature of the later phases of exponential growth. Thomas Malthus, whose predictions we met in the first chapter of this book, was just such a man, but he is only the most notable of a venerable line of what might be called "exponentialists"—a line that is still very far from extinct today.

In nature, though, systems whose growth is exponential do not pursue the steep upward curve very far. Bacteria kept in a state of optimum warmth on a plate of jelly nutrient, for instance, do not multiply indefinitely. For a while they double their numbers every few hours in classic exponential fashion; but when they have covered the entire surface of the jelly and there is nowhere for them to spread to, their numbers stabilize. The curve then bends over again and tends to flatten out, so that it comes to resemble a lazy "S" rather than the trajectory of a rocket

The Discovery of the Elements - An Example of Exponential Development

The Exponential Curve

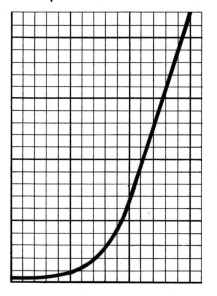

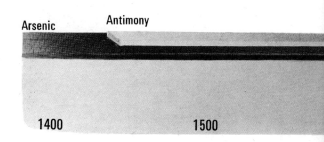

Arsenic

Antimony

1400

1500

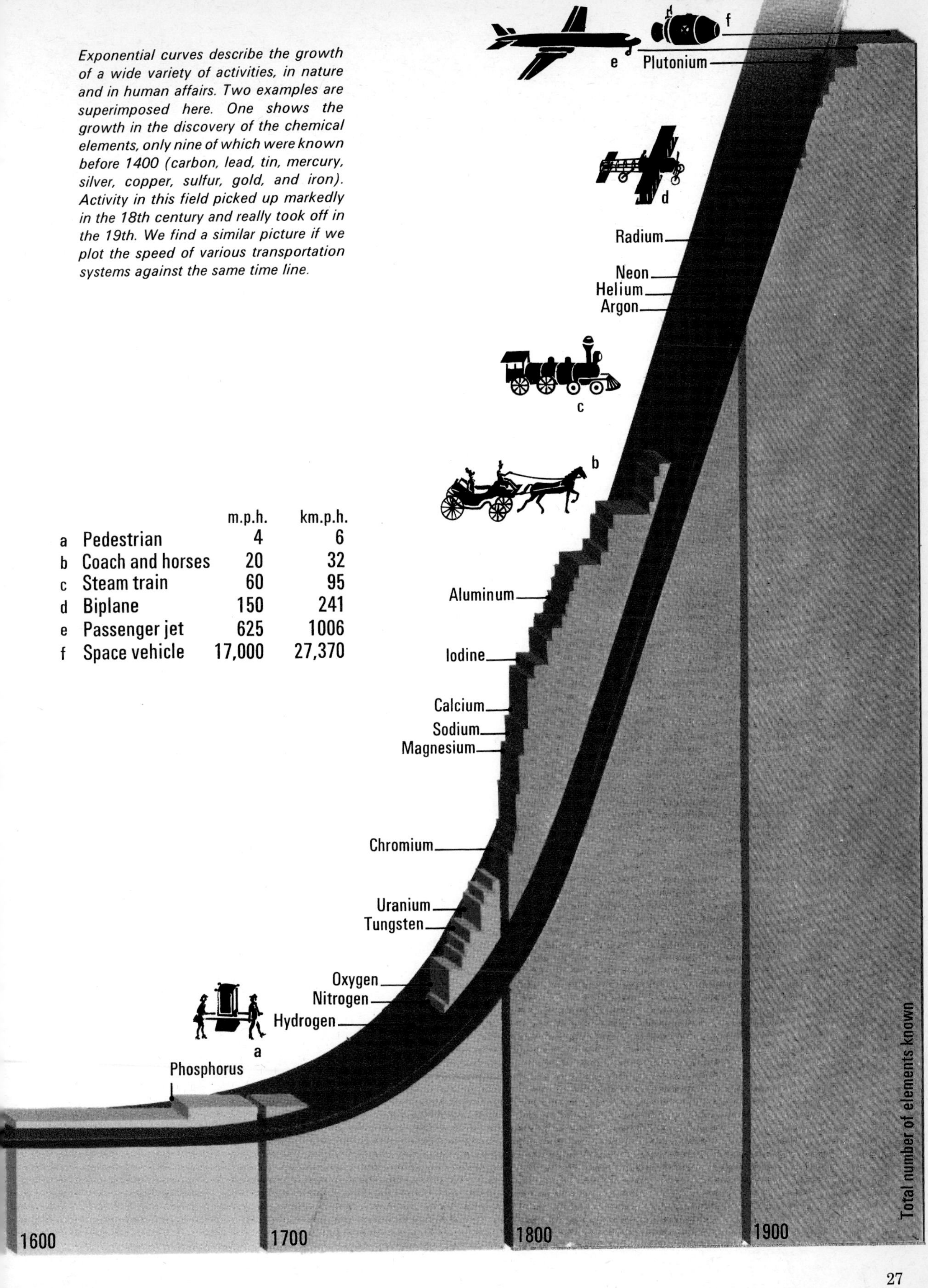

Exponential curves describe the growth of a wide variety of activities, in nature and in human affairs. Two examples are superimposed here. One shows the growth in the discovery of the chemical elements, only nine of which were known before 1400 (carbon, lead, tin, mercury, silver, copper, sulfur, gold, and iron). Activity in this field picked up markedly in the 18th century and really took off in the 19th. We find a similar picture if we plot the speed of various transportation systems against the same time line.

		m.p.h.	km.p.h.
a	Pedestrian	4	6
b	Coach and horses	20	32
c	Steam train	60	95
d	Biplane	150	241
e	Passenger jet	625	1006
f	Space vehicle	17,000	27,370

Plutonium

d f

e

Radium

Neon
Helium
Argon

c

b

Aluminum

Iodine

Calcium
Sodium
Magnesium

Chromium

Uranium
Tungsten

Oxygen
Nitrogen
Hydrogen

a

Phosphorus

Total number of elements known

1600   1700   1800   1900

into space. And when the bacteria have used up all the nutrient in the jelly, they begin to die. The curve that describes their decline is one that dips downward and heads back toward zero. It ends up looking like a camel's hump.

This camel's-hump curve describes the course of a large number of once-for-all activities. A forest fire, for instance, may start small, rise to an uncontrollable peak, then, as the trees and litter are consumed, sink to a smolder and eventual extinction. Or another example might be the emptying of a dammed-up reservoir through a small breach in the dam that widens as water flows out. In short, the camel's-hump curve describes what happens when a limited supply of something, such as burnable timber or penned-up water, is consumed exponentially.

Events that take a cyclic course move along entirely different curves. An obvious illustration is the temperature of a room heated by a thermostatically controlled heater. The temperature drops, the thermostat cuts in, the heater warms the room, the temperature rises, the thermostat cuts out, the heater stops, the temperature falls again . . . and so on. The curve that describes this pattern *oscillates*; that is, it rises and falls like a wave. Nature, too, has many examples of this kind of event. We can see it at its simplest whenever a new species invades a favorable territory, or is introduced into it deliberately, as the Australians introduced rabbits into their continent in the last century. At first the numbers of the new animal zoom upward exponentially, but then the fodder that permits such swift growth—grass, say—begins to give out, and the numbers are checked by starvation and disease. The decline, which may be drastic, permits the grass to recover, and so the grazers' numbers also recover in time. Thereafter, grass and grazer rise and fall in a regular way. They do not do so as regularly, however, as does the temperature in a room with thermostatic heat control, for too many other factors bear down on the natural system; weather, competition of other grazers, the pressure of carnivores, even the changes of the seasons—all such things help to distort what would otherwise be a sturdily regular oscillation.

But to distort is not to efface. The oscillations are still there. And with skill it is possible to use such curves for making worthwhile predictions.

To sum up, then, pure exponential increases are likely to happen only in the world of pure

mathematics. In the real world, as a general rule, quantities that begin by increasing exponentially either collapse on the camel-hump pattern or oscillate with a variable degree of regularity. In human affairs, the quantity that probably concerns more people than any other is the level of economic activity—even though various individuals might use different words to describe it. In personal terms, economics governs the answers to questions of where and how people live, what vacations (if any) they take, the food they buy, the leisure time they have and how they fill it, and the size of their savings or debts, to name only a few of the more obvious categories.

At the global level, economics determines the whole of mankind's impact on the earth; economics is, therefore, the greatest source of change currently operating. That explains why *any* forecasting of *any* particular future has to be at least partly based on some estimate of future economic levels. It explains, too, why most of the professional forecasters around the world are primarily engaged in economic prediction.

On the face of it, economic activity is ideally suited to forecasting by means of mathematical curves and the equations that govern them. Take the following sequence of events, for instance:

Suppose that in a given year people suddenly buy more than they bought the year before. Manufacturing output goes up. So profits go up. People earn more—through overtime, for instance, or higher wages, or increased dividends. At the end of the year, everyone is richer than he was the year before. So next year people buy even more than before, and the exponential curve looks about ready to take off.

This is a situation in which every apparent factor either directly or indirectly affects everything else. If the factors mentioned were the only ones at work, the situation would be ideally geared for runaway growth of the exponential kind. But there is another sequence of events that is working alongside the first: as manufacturing steps up and unemployment falls, wage rates rise and eat into profits; profits fall, cutting the amount of wealth to be shared around; and so, after all, people do not buy so much next year. This sequence, running directly counter to the other, checks its runaway tendencies, and the result is oscillation. Dozens of other factors, of course, are also involved: taxes, inventories, imports, interest rates, investment, government spending, and many more. Nevertheless, they are all related in essentially the same way as those in the two primary sequences already discussed.

Their complex relationships are not haphazard or capricious; they are, indeed, fairly regular—which is another way of saying that with extreme diligence you could find a mathematical equation that would express each of the numerous relationships. If you could find enough equations, you could in effect build a mathematical model of the entire economy.

The acknowledged master at that business is Lawrence Klein, who is Benjamin Franklin Professor of Economics at the University of Pennsylvania's Wharton School. He has been building and improving mathematical models of America's economy for over 30 years, and the "Wharton model" for which he is responsible is a highly respected technical aid to economic forecasting. It spells out the interrelationships among more than 170 variable quantities, all of which must be simultaneously determined in order to arrive at a forecast.

The Wharton model is only one of several highly thought of and commercially successful models. How accurate are they as compared with plain, old-fashioned "commonsense" judgment? In fact, neither the mathematical world nor plain judgment give very reliable predictions according to a recent study by America's Federal Reserve Bank. In 1974 the country's rate of economic growth actually declined 2.2 per cent, whereas the three most respected models had predicted increases of 1.2, 0.7, and 0.6 per cent. Similarly, prices rose 10.2 per cent, as against predictions of 6.5, 6.6, and 7.2 per cent. However, those economists who relied mainly on their good judgment did no better, either then or at any other time. (For an ironic sidelight on the forecasting business, consider the fact that a colleague of Klein's left Wharton School in 1969 in order to establish his own economic service. Klein was brilliant, he said, but was "a flop as a businessman." Nevertheless, the colleague himself unfortunately failed to foresee the full effects of the 1970 recession and had to sell out to a major bank in the following year!)

Klein maintains that the data presently being gathered even in advanced economies of the West are just too imprecise to permit truly accurate forecasting. Progress will come, he says, through better understanding of the structure of the economic system in general; equations are, after all, merely a mirror of the structure.

He is now seeking to push that understanding outward, to build, in cooperation with economists in several other countries, a model of the world economy, no less. It already consists of more than 1500 simultaneous equations—and it is by no means complete. Such efforts to improve our understanding of relationships within the economic system and the structure of the system as a whole are vital not only to our own future but to the future of life in general, because the two kinds of future are now so closely interwoven. But though we recognize the potential importance of mathematical forecasting, we are right to bear in mind its indifferent performance so far, and to maintain a proper skepticism when we consider—as we shall in the next chapter—even more ambitious attempts to build models for the whole of civilization not for just one or two years but for five or 10 *decades* ahead.

Any unbiased comparison between mathematical computer-based forecasting and the traditional human-judgment kind indicates that the human brain still has two great advantages over the machine. In the first place, the brain can make intuitive leaps that at present no imaginable computer will find possible; the brain gets to the root of things quicker and more surely. Secondly, it can *know* the real world. For example, it can identify a figure as impossibly high or impossibly low if one turns up as a result of its computations. A computer can never do that. Indeed, when, as often happens, an existing computer model produces a ludicrously unrealistic figure, the human beings who control the model must step in with a temporary program instructing the computer to ignore that particular result and to accept in its place a suggestion for a more reasonable number.

The unique combination of intuition and common sense makes the informed human brain a powerful forecasting "machine" in its own right—even, as the Federal Reserve Bank study showed, in the highly mathematical realm of economic forecasting. In many other fields of endeavor, where no strict quantity relationships exist among the various component elements or decisions, the brain is the only forecasting machine worth considering. A typical such field is medicine, where an advance in heart surgery, say, may have immediate effects on, for instance, kidney surgery—or may have none at all. In the field of economics, it is hard to find a parallel case where the interactions are so indeterminate,

Two Economic Trends that Balance Each Other Out

People buy goods, and this inevitably encourages further manufacture.

So output goes up—and with it go the wages that are earned by the workers.

But wait! As output rises, unemployment falls. Wage rates go up as spare workers grow scarce in the market.

Rising output usually leads to rising profit, so there are greater dividends.

Fatter wage packets eat into the profits, so that new wealth, instead of increasing, actually starts to decrease.

Workers get more money, too, as they work over-time to meet demand, so everyone has more money.

These two sequences summarize two basic processes at work in any free economy—the up-and-down relationship between the money people spend, the goods they buy, industry's response to that demand, and the new wealth that results from such industrial activity. They show how, even in the same sequence of events, there can be two quite contrary tendencies, one acting so as to accelerate, and the other to decelerate, economic activity. In every economy there are literally hundreds of similar tendencies, some with a positive effect, others with a negative one.

With less money to go around, demand for goods falls. Production follows demand down, and thus everyone ends up with less money.

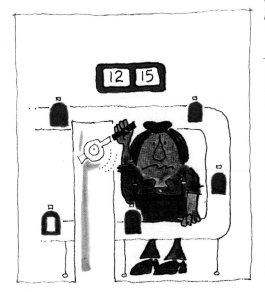

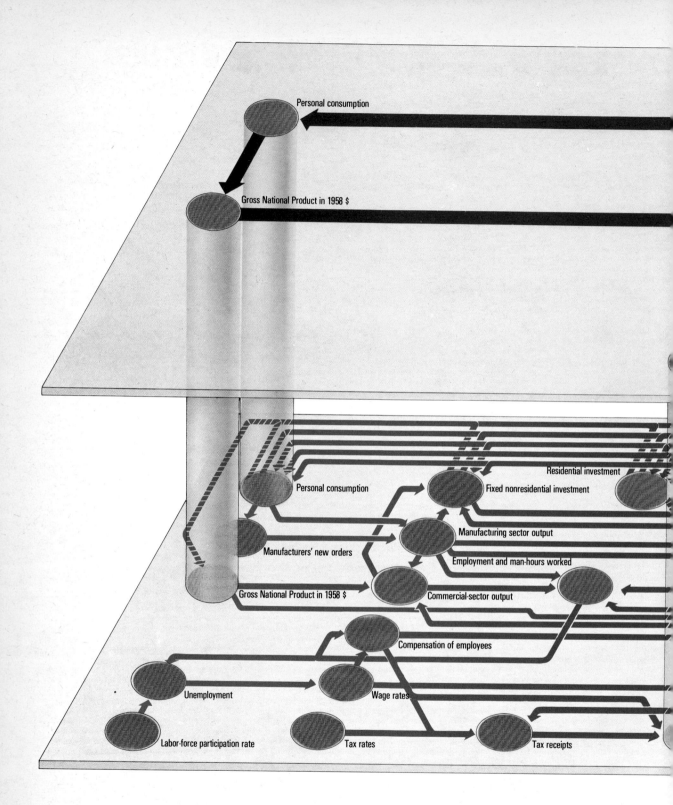

Personal consumption

Gross National Product in 1958 $

Residential investment

Personal consumption

Fixed nonresidential investment

Manufacturing sector output

Manufacturers' new orders

Employment and man-hours worked

Gross National Product in 1958 $

Commercial-sector output

Compensation of employees

Unemployment

Wage rates

Labor-force participation rate

Tax rates

Tax receipts

The Wharton Model of the US Economy

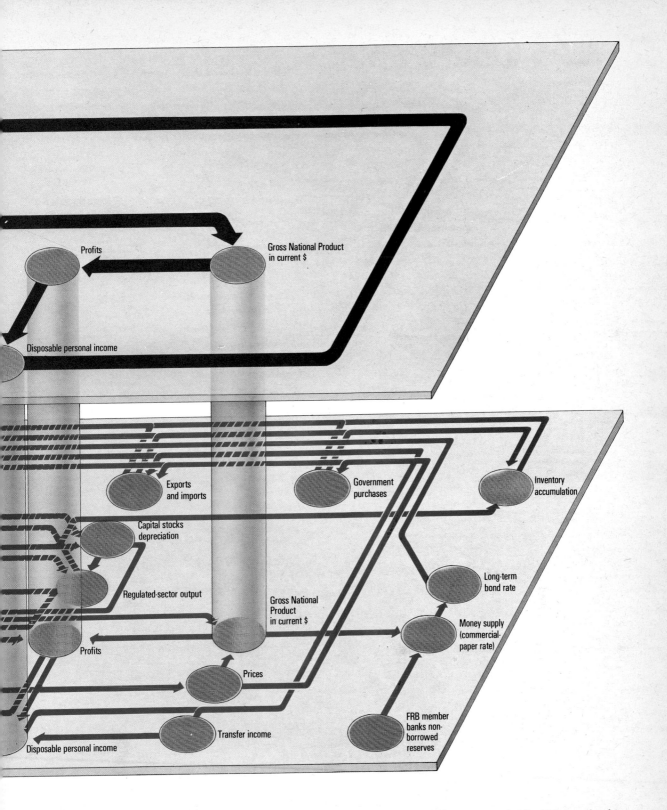

Profits

Gross National Product
in current $

Disposable personal income

Exports
and imports

Government
purchases

Inventory
accumulation

Capital stocks
depreciation

Long-term
bond rate

Regulated-sector output

Gross National
Product
in current $

Money supply
(commercial-
paper rate)

Profits

Prices

Transfer income

FRB member
banks non-
borrowed
reserves

Disposable personal income

The lower half of this two-part chart shows the main elements that interact to keep the US economy going (models for other free-enterprise economies would appear very similar). In this diagram an arrow merely indicates that one of the variables has an influence on one or more of the other variables (e.g. rising prices may increase gross national product but decrease exports). Each such arrow in the diagram is represented by one or more simultaneous equations in the computer. The daunting complexity of this sort of model soon becomes apparent. The diagram in the upper half extracts just five elements from the main Wharton model of the US economy and shows how these elements are related to one another in an endless cycle.

but the relationships in *most* human activities are precisely of that loose-structured kind. Mathematical forecasting has no value there. For any sort of meaningful prediction, we have to turn to that other tool, the informed human brain. We can all do it; there is no special gift of prophecy involved. Try this forecast:

You have before you the personal history of a man who is about to take the next big step in his life. He began as an apprentice in metal engineering. When he qualified, he helped install some machines in a plastics plant, way back in the early days of plastics. The new materials fascinated him, and he pioneered a way to form plastic tubing. The sales office decided to send him rather than a professional salesman to clinch the first deal on this product, because there was no time for sales people to bone up on the basic science that the new process involved. It was a wise decision. He could talk to clients' engineers in their own language. And he certainly knew about metals—the competition.

Soon he was working full-time in sales. Then he headed a new department: sales development. A conglomerate took over the firm and moved him up to head their central marketing division. He hoped for a managing directorship next, and the offer came—not from his own company but from its biggest competitor. There was a blazing row with his chairman, but he took the job. For the last seven years he has held it more successfully than even he dared to hope. Now he is about to make the next big move of his life. What is it?

Chairman of the group that lured him away? That would certainly be the least surprising step.

Or chairman of the old conglomerate? It would perhaps be more surprising, but by no means an unlikely move.

Or does he become a warden in a national park? This would appear to be an extraordinary move—unless you happen to know that his doctor has given him the gravest possible warning about what corporate life has been doing to him. That being so, the decision would not be in

No existing computer, and none in the foreseeable future, can match the human brain for its intuitive leaps of imagination and inventive genius. That particular genius can show very early in life, as witness the "park for excercising dogs" redrawn opposite from a design by a 9-year-old girl. Below: a scene from the film 2001: A Space Odyssey. *The film included a warning against letting computers take over our affairs.*

After Edward de Bono,
The Dog Exercising Machine,
Jonathan Cape Ltd., London, and Simon and Schuster, Inc., New York

the least surprising. One extra nugget of information, and the whole picture changes!

Or does he become a world-famous concert violinist? Nonsense! We know that fact is often stranger than fiction, but there really is no way he could make such a change.

There are no hard and fast equations to govern any of the above probabilities except for the last, where probability = zero is an easy one to write. For the others, estimates (or "guesstimates") such as "highly probable" or conditional estimates such as "improbable unless . . ." are more in order. To assume such kinds of probability the human mind is an amazing computer. It sees patterns almost before it grasps anything else; and where there is no pattern it will even invent one—as it invented a Great Bear, and Orion, and the Seven Sisters to reduce the chaos of the stars in the night sky to a manageable order.

This compulsion to find a pattern is the basis of a number of nonmathematical forecasting techniques. The most widely practiced is the Delphic-study technique, named (with more self-putting-down humor than logic) for the famous oracle at Delphi in ancient Greece. This technique was formalized in the early 1960s, but it had its origins in the previous decade in a much cruder technique known as "brainstorming." The story of its development is itself revealing.

During the 1950s it became clear that the world was changing faster than even the most visionary prophets of the 20-odd years before the outbreak of World War II had thought possible. Habits of thought and expectations formed in earlier, slower times were now a hindrance to judgment when questions of research, investment, and other forward-planning moves were being considered. What was needed was a technique for shaking up the old habits—hence "brainstorming." At a typical session of brainstormers, the experts sat around a table and encouraged each other to indulge in wild flights of fancy in their own and related fields. Practicalities could come later. For the moment, all that mattered was that no possibility, however far-out, should go unrecorded. And wild ideas such as color TV, mass tourism, the 120-member United Nations, quadraphonic sound, moon landings, megaton bombs, organ transplantation, Chinese-Russian discord, Black Power, and Unisex clothing, were all suggested as feasibilities. Also, to be sure, were

You could define humans as pattern-making and pattern-seeking animals. Give us an area of smooth sand (left) or a large garden (above) and we feel compelled to inscribe some sort of pattern upon it. That same obsession with patterns makes the human mind an extremely useful tool for picking out likely future trends.

others such as nuclear holocaust, reversion to barbarism, cheap nuclear power, the mile-high city, rapid mass transit, plastic lamp filaments, and domestic robots. The big problem was how to sort out the probables from the possibles, and both from the unlikelies?

One great disadvantage of brainstorming is that in order to develop a few new ideas, you really need to have a decision group of between 40 and 70 experts—an impossible number to manage informally. A second drawback is that once a person has exposed an idea publicly, he too often feels impelled to stick to it and defend it. And so brainstorming eventually gave way to the Delphic-study technique. Delphi gets around the difficulties by being, in essence, a brainstorming session that is conducted by letter and questionnaire and that lasts several months.

The typical Delphic study has three phases. In the first, a limited number of experts are canvassed about the aims of the study and are asked to suggest fruitful areas for inquiry and a period of time that predictions might reasonably cover. In the second phase, a much bigger panel of experts is invited to make predictions about the events suggested in phase one, and to add their own suggestions for related areas of interest that, in their view, have been overlooked. In the final phase, the same large panel is sent the earlier results and asked to make final predictions after a consideration of all the material in hand. Only the administrative researcher knows which of the experts said what in phase two, and so nobody has made an identifiable public commitment that he feels he must defend for the sake of his reputation. In most such Delphic projects, the participants are now asked to specify a year when they think a given event has a 50-50 chance of happening and a later year when, in their opinion, it is almost certain to have happened. The useful time scale for the technique is no more than half a century into the future—about as far ahead as most informed realists are willing to think seriously and make serious forecasts.

The real purpose of a Delphic study is not so much to link hard-and-fast dates to a number of likely events as to open our eyes to a likely sequence. Once we see the probable sequence, we can look at its deeper implications and take precautions to cushion ourselves against nasty shocks. In other words, Delphi helps with large-scale and long-term planning more than with the precise allocation of next year's budget and resources.

Even so, the technique has defects. Its chief flaw has been that it underplays the relationships among predicted events. For instance, the consensus in a Delphic medical study might anticipate trouble-free heart transplantation, even from animals to man, in the late 1980s—and might at the same time predict that a mechanical heart replacement will be perfected in the early 1990s. A moment's thought shows that once it becomes possible to put cheap animal hearts into people, there will be no further point in spending money for mechanical substitutes. Every technical advance affects the chances for dozens of other technical advances, some of them very remote. Who, for example, would have thought that the development of polytetrafluorethylene (PTFE) as a solid-lubricant coating for moving parts in deep space, where liquid lubricants would evaporate, would make for a revolution in non-stick kitchenware and woodworking tools? Well, strange though it seems, it did.

Sometimes, in fact, technology in a certain field takes a freakish leap forward quite independently of technical advances in related fields. Radar is a good example; it grew out of something that seemed a technical problem rather than an advance: the infuriating tendency of TV signals to bounce off buildings and other big solids and cause interference patterns on the screens of pioneer viewers in the 1930s.

There is probably no way to predict freak effects and freak advances such as radar. But another kind of unexpected event—the spin-off or cross-fertilization—can to some extent be allowed for. Recent Delphic studies have further reinforced their technique by adding a fourth phase in which the panel is asked to estimate the impact of one predicted development on another. The results, properly tabulated, form what is awesomely called a "cross-impact-matrix-forecast."

Whatever the forecasting technique of a modern futurologist may be, he tends to express his predictions—or those of any group he represents—in the form of a "scenario." This term, which is actually an Italian word for a preliminary outline of the plot of a drama before the script is written, was adopted in the 1960s by the futurologists. It aptly describes the plot-like shape in which they generally flesh out the bare-bone results of their studies. A good analogy would be the traditional manner in which a military commander expresses his view of a battle situation. As an example, let us take the famous discussion between Brutus and Cassius in Shakespeare's *Julius Caesar*, when they are considering whether or not to march their army

In the past it was enough for a development to be technically possible for someone to try it out somewhere. Nowadays a wider range of criteria, including social acceptability, determines whether or not we carry out such projects. Mile-high cities (right), possible since the 1960s, are a case in point.

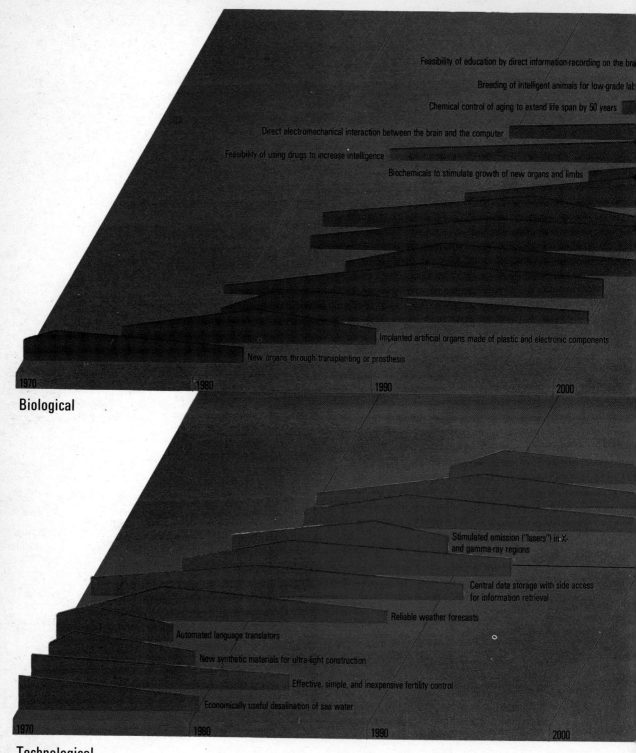

Feasibility of education by direct information-recording on the bra

Breeding of intelligent animals for low-grade lab

Chemical control of aging to extend life span by 50 years

Direct electromechanical interaction between the brain and the computer

Feasibility of using drugs to increase intelligence

Biochemicals to stimulate growth of new organs and limbs

Implanted artificial organs made of plastic and electronic components

New organs through transplanting or prosthesis

1970   1980   1990   2000

Biological

Stimulated emission ("lasers") in X-
and gamma-ray regions

Central data storage with side access
for information retrieval

Reliable weather forecasts

Automated language translators

New synthetic materials for ultra-light construction

Effective, simple, and inexpensive fertility control

Economically useful desalination of sea water

1970   1980   1990   2000

Technological

A Delphic Study of the Future of Technology

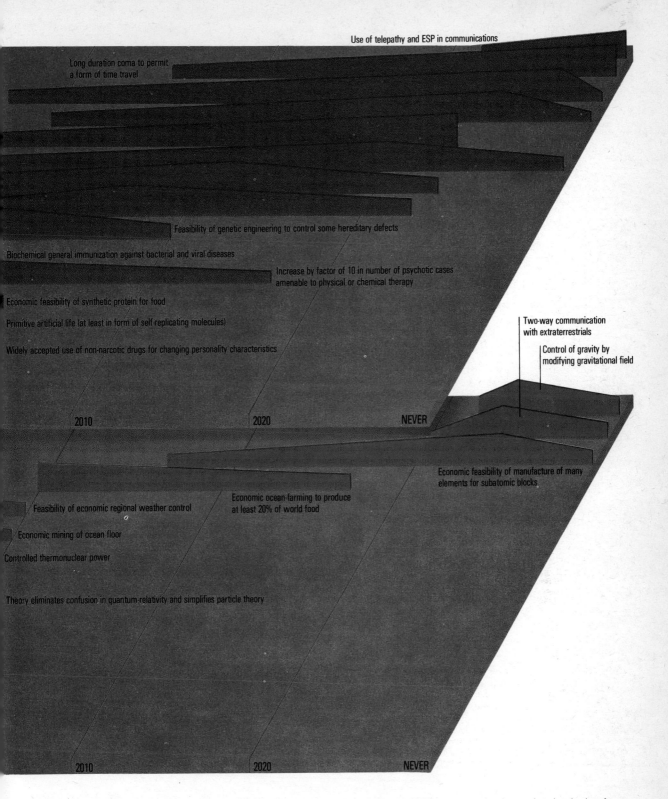

Use of telepathy and ESP in communications

Long duration coma to permit
a form of time travel

Feasibility of genetic engineering to control some hereditary defects

Biochemical general immunization against bacterial and viral diseases

Increase by factor of 10 in number of psychotic cases
amenable to physical or chemical therapy

Economic feasibility of synthetic protein for food

Primitive artificial life (at least in form of self-replicating molecules)

Widely accepted use of non-narcotic drugs for changing personality characteristics

Two-way communication
with extraterrestrials

Control of gravity by
modifying gravitational field

2010   2020   NEVER

Economic feasibility of manufacture of many
elements for subatomic blocks

Economic ocean-farming to produce
at least 20% of world food

Feasibility of economic regional weather control

Economic mining of ocean floor

Controlled thermonuclear power

Theory eliminates confusion in quantum-relativity and simplifies particle theory

2010   2020   NEVER

These diagrams show the results of an early Delphic Study carried out in the mid-1960s among scientists and technologists from a wide spectrum of disciplines. They were canvassed for their opinion as to when each of 31 breakthroughs was most likely to occur. (They were also asked to indicate their degree of knowledge about and involvement in the field in question, so that a due weight could be given to their opinion.) The preliminary results were then summarized and passed back to the experts for a more considered assessment. It is that more considered assessment that is illustrated here. The high point on each bar shows when half the experts thought a given development would occur; the cutoff shows when 90 per cent agreed on its likelihood.

41

from Sardis, in what is now called Turkey, to meet the enemy at Philippi in Macedonia, Greece, several hundred miles away.

Cassius's scenario is less energetic than Brutus's. We stay here, he says, resting, exercising, looking to our defenses, preparing to fight. The enemy has to seek us out, and he will waste his strength and wear out his men before the battle starts. But Brutus has a different vision of what should happen. We must go at once to Philippi, he says. The people between here and there favor the enemy's cause, not ours. At the moment we are at the peak of our strength. But if we let the enemy come to us, they will pick up recruits on the way, and we could easily lose our momentum. We are ready now; they are not. So let us go out and take them.

Soldiers and people in commerce and government have always found a use for this kind of forecasting based on a consideration of the likely results of alternative courses of action. The difference today is that the various departments of human activity are much more interconnected than they used to be. Thus it now

Left: this simple apparatus, a hair drier, a coffee can, and part of a kitchen scales, was enough to convince Britain's Christopher Cockerell of the feasibility of the hovercraft (shown below). But enormous labor and expenditure separated it from the finished project, which nearly died several times. History seems inevitable only to those who come to it with the advantage of hindsight.

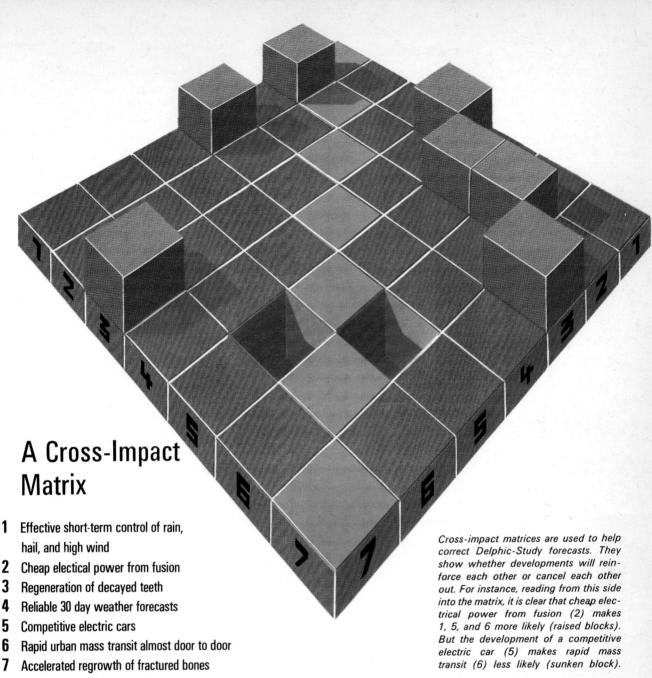

A Cross-Impact Matrix

1 Effective short-term control of rain, hail, and high wind

2 Cheap electical power from fusion

3 Regeneration of decayed teeth

4 Reliable 30 day weather forecasts

5 Competitive electric cars

6 Rapid urban mass transit almost door to door

7 Accelerated regrowth of fractured bones

Cross-impact matrices are used to help correct Delphic-Study forecasts. They show whether developments will reinforce each other or cancel each other out. For instance, reading from this side into the matrix, it is clear that cheap electrical power from fusion (2) makes 1, 5, and 6 more likely (raised blocks). But the development of a competitive electric car (5) makes rapid mass transit (6) less likely (sunken block).

takes very special skills—and the expertise of many people—to prepare a useful scenario. Even to work up a scenario for one restricted field—for transportation, say—may call for the combined expertise of engineers, town planners, social psychologists, statisticians, demographers, economists, construction firms, research laboratories, politicians, pollsters, and many others.

The name most closely associated with the scenario-making branch of futurology is undoubtedly that of Herman Kahn, formerly of The Rand Corporation, later of the Hudson Institute (both of which specialize in analyzing

trends and planning for the future). Kahn, together with Anthony J. Weiner, compiled the monumental *The Year 2000*, which was published in 1967. It was not so much a book of predictions as a sort of teach-in on how to think about the future—"a framework for speculation," to quote its subtitle. It contains dozens of scenarios, all of which, when taken together, show us what a very frighteningly rich network of possibilities the unknown future contains.

Even so, despite that richness, it is illuminating to leaf through the book today, a decade later, and to realize that not one scenario foresees any one of the following events: the discovery of vast petroleum reserves on the European continental shelf; the quadrupling of crude-oil prices within a few months; worldwide inflation between 6 and 20 per cent; the spontaneous overthrow of Portugal's long-established right-wing dictatorship and the voluntary dismantling of her colonial administration; Sino-American détente and US backing for communist China in the United Nations; and the enforced resignation of an American President. Clearly, the actual future of the world in 1967 (the part of the future that has now become history) has already deviated in several major respects from the hypothetical future as it appeared to a number of intelligent, perceptive, dedicated, full-time futurologists, who had all the "think-tank" backing they could possibly use.

The unpredicted events just listed have one thing in common: they were surprises. Nothing that had happened up to 1967 had prepared people to expect such developments. This highlights a major disadvantage of working with experts. They are very good at dealing with hard facts, but experience shows that they are *not* good at imagining the surprises that lie ahead, especially in their own fields.

After all, it was not an experimental scientist but a fantasist and spinner of tales, H. G. Wells, who foresaw men on the moon and atom bombs half a century before they happened. It *was* an eminent scientist (Britain's Astronomer Royal, no less) who, with his full understanding of all the problems involved, announced in 1956—just a year before Sputnik 1 went into orbit—that any idea of space travel was "utter bilge." Learned journals from the first 50 years of this century are embarrassingly full of articles by great men, proving with absolute mathematical certainty that space flight is impossible, that no large passenger aircraft would ever cross the Atlantic nonstop, that the extraction of nuclear

The fantasy merchant and spinner of tales is often a better guide to the future than the most eminent expert. Jules Verne's From the Earth to the Moon *was first published in 1865. It correctly foresaw that the return of the moon vehicle would culminate in an ocean splashdown. Verne even foresaw that the vehicle would be American, although in 1865 America was hardly the world's leading industrial power. The actual event (shown opposite) took place in July 1969, more than a century later, yet the similarities between the two pictures are startling.*

energy was impossible, and so on. Lord Rutherford, who had done as much as any man to reveal the nature of the atom and of the vast forces that hold it together, remained wittily scathing to the end of his days about anyone who dared to predict useful nuclear energy. And yet in 1942, only five years after Rutherford's death, the first controlled, energy-yielding chain reaction started up in some make-shift apparatus put together on a tennis court in Chicago.

Even to have been right in the past is no guarantee of continuing imaginative success. Edison and Swann made the electric light bulb when every other respectable scientist could easily prove how impossible it was; yet Edison's long-fought rearguard action against alternating-current supply substantially held back electrical engineering in the United States. Expert knowledge, it seems, can clog the imagination as easily as stimulate it. In fact, technical knowledge

may not be needed at all. What do you think of the following prediction—and when do you think that it was written?

"Devices may be made by which the largest ships, guided by a single man, will travel faster than if they were full of sailors. Vehicles may be built to move at great speed without draft animals. Flying machines may be constructed in which men, sitting comfortably and entertaining themselves, may wing their way through the skies like birds . . . also machines to enable man to explore the very depths of the oceans. . . ."

The diction has certainly been modernized ("devices" for "instruments," "vehicles" for "chariots," and so on), but the ideas are precisely as the English Franciscan Roger Bacon expressed them in the 13th century.

The lesson must be that any forecast assuming a future that is continuous with the present is probably wrong, and possibly badly wrong. We

Left: a French book of 1887, La Vie Electrique *(Electric Life), correctly foresaw that electric power would in some ways liberate mankind, but would unleash a host of new problems too. Below: today, with nuclear power, we face the same sort of dilemma.*

Above are pictured two recent developments that would surprise (and probably shock and even horrify) the average Westerner of the 1920s if we could go back in time and reveal them—in other words, these developments were foreseen. One is a mass pop festival of the Woodstock kind, though this one actually took place in Rotterdam, in the Netherlands. The life style and behavior of the fans would be incomprehensible to the 1920s generation. So, too, would be the current enthusiasm for oriental religions as exampled by London's Hari Krishna sect. By contrast, World War 2 (the Dunkirk scene shown at right is from a painting by Charles Cundall) despite its vast scale and its new technology, contained very little that would have surprised our hypothetical person of the 1920s. It was a straight development from already well-established trends.

do not live in surprise-free times—and surprises make for discontinuities. Just work back over the past three quarters of a century and think of the developments that have taken us by surprise, compared with some that were foreshadowed and expected long before their time came. Here are a dozen of each, jotted down at random.

Unforeseen
Plastics
Pop culture
"Modern" art
Mass air travel
Ball-point pens
Computers
Westernized Oriental religions
Population explosion
Radio media
Nuclear diplomacy
Electronics
Hollywood

Foreseen
Colonial freedom
Television
Cars
Therapeutic drugs
Space flight
Selective pesticides
Female equality
Decline of traditional religions
Increasing wealth
World War II
European economic federation
Leveling of racial barriers

The unexpected events were, in effect, invisible obstacles to useful prediction. They were like wild cards in the pack. The extraordinary thing, though, is that even the items in the "foreseen" list have had effects that no one predicted, and have failed to have effects that everyone confidently expected. Television, for instance, was going to restore the family circle to its Victorian glory; it was going to kill the printed book and the movies; it was going to make classroom teaching obsolete. So far, at least, it has fully accomplished none of those ends.

In short, failures and misreadings haunt the whole futures industry.

Alternatives for Man

It may seem arrogant to consider the future of man before we turn to the future of life in general. In fact, though, man's potential for good or ill on the face of the earth is now so great that the future of billions of creatures, representing perhaps millions of species, probably depends more on our decisions and our level of activity than on any other factor. Few people can be ignorant of that dependence. One of the most explosive growth industries of recent years has been the production of books, articles, documentaries, and lectures on futurology, population, eco-doom, and related topics. Nothing recycles more efficiently than a disturbing fact about 20th-century technology or modern civilization.

It was not always so; only a couple of decades ago we were vastly more optimistic. Looking back now, from the standpoint of the late 1970s, it seems beyond belief that as recently as the 1950s it was almost universally believed, outside the Communist-dominated countries, that human history had reached its culmination in European-American society. Europe-America had almost conquered inequality, poverty, disease, and all but the natural disasters. Within acceptable limits we were managing our economies well. Obviously, no further structural changes were needed—just a little mopping up here and there where things were not yet quite perfect. The once-great bogey that socialist revolution would arise spontaneously within our industrial society, a bogey that had haunted capitalism for almost a century, held no more terror for us.

"America today, tomorrow the world," we used to say, and a large part of the world agreed. The American Dream was a world dream, spelled out in a widely quoted prophecy made by Philip Handler, former President of the US National Academy of Sciences. In the future, he said, "the bulk of humanity will be gathered in megalopolises, dwelling in huge buildings surrounded

Not so long ago the vast African grasslands must have seemed secure from drastic human intervention at least during the present century. Yet already we are talking of emergency steps to preserve even the national parks and game reserves from overexploitation and encroachment—areas that were specifically set aside to maintain animals in their natural wilderness.

Left: the TV-phone has figured in most visions of the technologically advanced future. At present it exists only as an inter-office communicator, but there is no technical reason to prevent its going into general telecommunications service.

Below: the domed-over, air-conditioned city is another vision common to many futuristic projections made in the earlier years of the century. Today, more cost-conscious and less awed by sheer technology, we should certainly discount its possibility in the foreseeable future.

by parklands, partly covered with domes within which the atmosphere will be maintained rather constant. . . . Each individual will have a private, pocket two-way television instrument and immediate personal access to a computer serving as his news source, privately programed educational medium, memory and personal communicator with the world at large—his bank, broker, government agents, shopping services, etc. . . . The bulk of the labor force, then, will engage in activities currently classified as services rather than production of goods. . . . Most of us hold such a dream in common. . . . The most important thing one can say about that dream is that it may well be feasible." In a different context, Handler emphasized strongly the fact that science "*is* capable of fulfilment of our dream."

Increasingly through the succeeding years, as the less anticipated effects of high technology—ranging from electronic surveillance through ecological blunders—have surfaced, the technocratic vision has dimmed considerably. To many people, it is the future alternatives to our menacing present that have begun to look attractively rosy. "*Alternative*" has become a key word in the vocabulary of the 1970s.

To be sure, there always were people for whom the dream of a scientific megalopolis—the "megadream," we might call it—was more nearly a nightmare. But they tended to be nonscientists.

One vision of the good life has come true for a fortunate few of the world's teeming millions, as, for instance, for the owners of these individually styled and lavishly equipped homes in San Diego, California (left), each of which has its own private pool. People who live at such a standard (and they number into the millions in the West alone) consume the world's resources of energy and materials at least 600 times faster than the world's poorest folk (who number billions), here typified by the poor in Calcutta's decaying streets (right). The great questions facing us are: can the entire world population ever consume energy and materials at Western standards, or will such inequalities persist forever, or is the Western achievement freakish and temporary?

Certainly their challenge came from outside the ranks of scientific orthodoxy, and their vision of the future owed much to the past—to a supposedly simple life of the horse-and-buggy kind. Still, they were not exaggerating when they saw the megadream as a nightmare. Horror is implicit in the internal contradictions of the megadream itself: the fact that too many people are clamoring for their share of the "good" life; the fact that our known and likely-to-be-discovered resources will not go around; the fact that industrialization, which makes the megadream possible, cannot be applied on a world scale without inflicting intolerable damage on the rest of the living world; and the fact that even an apparently limited degree of damage to the whole environment might bring on our own extinction.

These facts are derived from studying present trends and projecting them forward into the near future. Let us look a little more deeply into the major problems that are at the heart of our present disquiet. The first of these, of course, is the current population explosion.

Early man, way back near his origins on the African plain, numbered perhaps 125,000 individuals before beginning to spread out to other lands and to exterminate other manlike competitors. Despite disease, famine, ice ages, and other hazards, mankind ultimately colonized all parts of the globe. By around 8000 B.C. the number of human beings approached 5 million. Then agriculture was invented, and things began to move. The human population of the world doubled, and redoubled, and redoubled . . . almost six times in succession, and it had reached 250 million by the beginning of the Christian era.

The Population Explosion

1400 AD
350 million people

1550 AD
400 million people

1650 AD
500 million people

1850 AD
1000 million people

1971 AD
3670 million people

2000 AD
7400 million people?

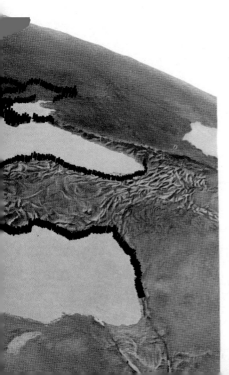

The time when numbers double is a significant figure for anyone who wants to grasp what population growth means in everyday terms, because it measures the period over which *everything* has to double—houses, transportation, educational facilities, invested and surplus wealth, and food supplies—if the situation is merely to stay stable. If people are to hope for some improvement, they must achieve it over and above the rate at which everything doubles.

The next doubling after Christ's time took a bit more than one and a half millennia. By A.D. 1650 there were 500 million people. It took only 200 years, to 1850, for the figure to reach 1000 million; and the number doubled to 2000 million in 1930, just 80 years later. You might think that the pattern 1600: 200: 80 was enough to alarm anybody, but in fact it alarmed very few. One person who *was* worried by it was the British demographer Sir George Knibbs, who in 1928 calculated a world population of 3900 million (almost doubling) by 2008—an annual growth rate of 1 per cent as compared with the 0.9 per cent increase from 1850 to 1930. His contemporaries treated his figures, with their implied warning, as a symptom of senile decay. The noted biologist Julian Huxley, speaking over the British radio in the early 1930s, also warned of population increase, but in more graphic terms, and suggested birth control as the only remedy. The BBC network's staunchly religious head, Sir John Reith, reprimanded him for "polluting the ether" with such tales. Even in 1950, when Huxley and other demographers warned of a population of 3000 million by the year 2000 (a more modest forecast than Knibbs's), they were accused of alarmist talk. In fact, it took less than 10 years after 1950 for the total population figure to pass the 3000 million mark.

In 1964, the United Nations Organization produced a set of very carefully projected forecasts of world population growth. Country by country, UN demographers examined birth, death, fertility, and female-replacement rates—every factor that determines increase—and made three provisional predictions. One, called "low," based on the assumption that family-planning programs would make some real progress in the coming years, foresaw a population of 5300 million in the year 2000. A second, "high" forecast predicted 7400 million if there were no progress at all in family planning. The "median" forecast was for 5800 million. In less than 10 years,

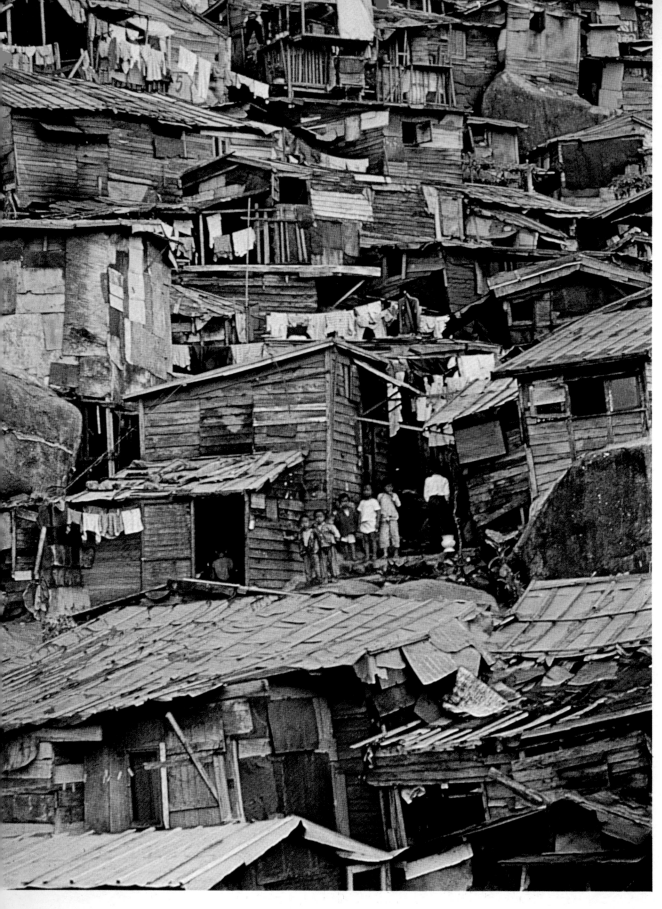

however, it became clear that the actual world population was moving toward a figure above even the UN's pessimistic "high."

It would seem that underestimations are the rule in population forecasting. Even those who want to warn the rest of us—and who therefore lean toward publicizing the most alarming sets of figures—usually underestimate the true rate of increase. The current rate of world increase in population has gone up from an annual 0.9 per cent over the period from 1850 to 1930 to just above 2 per cent today. This means that there will be a doubling within 33 years.

The situation seems especially alarming when one considers that the highest population-growth rates are being experienced in the regions that can least afford them. Central America has just 21 years in which to double its food supply, houses, schools, clinics, transportation capacity, gross national product—everything!—merely to maintain today's miserable standards for a probably redoubled population. North Africa has 23 years, southwest Asia and tropical South America have 24 each, southeastern Asia has 25, eastern Africa 26, western Africa 27, southern Africa 29, and central Africa 32. When you think that even the richest country, with the most solid economic base, could not seriously plan now to double its whole economy and all its amenities in a mere 30 years, what possible hope is there

For most experts overpopulation is the world's most pressing problem (below). Already most of the world's poor live in desperate conditions (left). To bring relief calls for massive publicity backed by spending of space-program proportions.

for these wretchedly poor regions of the world?

Even America, with 63 years to go, and western Europe, with 117 years, face an apparently impossible task, for the amenities that they must double are, in capital-per-head terms, 200 to 1000 times greater than those of poor countries. Thus, any discussion of population is bound to lead very quickly to the question of resources.

Let us forget for a moment all question of improving living standards in the already developed world. Let us simply take world population as it will be in the year 2000 (optimistically assuming that it will total just over 7000 million) and see how much of the commonly used metals would be needed merely in order to reach a standard of living comparable to that now enjoyed in America and Europe. To do this would take 60,000 million tons of iron, twice the quantity of our known profitable reserves. It would take 1000 million tons of copper; today's reserves are 200 million

Resources: The Emptying Storehouse

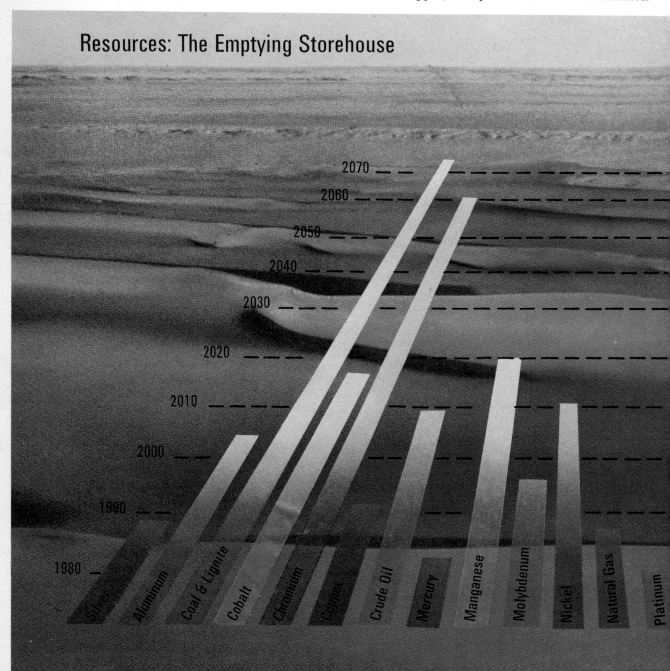

2070
2060
2050
2040
2030
2020
2010
2000
1990
1980

Silver
Aluminum
Coal & Lignite
Cobalt
Chromium
Copper
Crude Oil
Mercury
Manganese
Molybdenum
Nickel
Natural Gas
Platinum

tons. Where would the necessary 600 million tons of zinc come from, when our known reserves are only 85 million tons? And our reserves of tin are a mere 6 million tons—nowhere near the 100 million tons we should need. All that, remember, just to bring the world up to what we call a "developed" standard!

The above figures apply only to a handful of essential metals, but the same holds for cobalt, lead, manganese, mercury, molybdenum, nickel, and tungsten; all of these substances are patchily distributed, and all exist in very inadequate reserves of economic ore, with the uneconomic ores lying far beyond the reach of any acceptable mining costs. And for each of them there is at least one vital commercial application for which there is no conceivable substitute.

What is true of metals is true of other resources. Planet earth has vast supplies of all the things we use, but only the smallest fractions of them are economically winnable. Even if we had all the material resources we needed to satisfy the industrial requirements of the world of A.D. 2000, however, we have already lost one critical battle in the fight to save mankind from a miserable future. We have already lost the battle to produce enough food to go around.

Nothing is harder to pin down or to summarize briefly than the world's food balance. There is no precise agreement, for instance, on the number of calories a healthy person needs: it varies with stature, climate, age, occupation, and dietary constituents. The permutations among carbohydrate, fat, and protein are so enormous that even an agreed calorie scale would be subject to many qualifications. On top of that, the world's statistics-gathering agencies are better geared to measure production than consumption, and so caloric losses in the transportation, storage, and marketing of foodstuffs are often underestimated. They are known to vary between 10 and 50 per cent of production. Finally, the effects of dietetic ignorance and harmful cooking can turn an adequate supply of raw food into an inadequate diet. It is hard to take account of all these factors in such a way as to be sure that the resulting picture is anything like reality.

Nevertheless, even if we make all possible optimistic assumptions, so that total wastage through the various factors is only 10 per cent, the picture remains grim. To put it all into figures would make for a fairly unreadable page or two, but the accompanying diagrams sum it up. They show that Europe (including European Russia), North America, and Oceania (Australia, New Zealand, and nearby islands) are the only

Uranium Tungsten Zinc Gold Iron Lead Tin

The gloomiest projections of our future reserves of metals (those recoverable with present-day technology) foresee most of them running out before the middle of the next century, some well before that. Each of these metals has at least one application for which no other metal can possibly substitute.

regions where our current diets are more than adequate. Moreover—as an illustration of the extent of our failure to win the battle for food—if all the surplus enjoyed by all the consumers in the well-fed areas were taken away and redistributed among the hungry so as to give everyone in the world an equal plateful, everyone in the world would then starve. It would not be the sort of starvation that tugs at the heartstrings and stirs the conscience. It would be just the ordinary kind of undramatic hunger now endured by about 2500 million people every day of their lives—the kind that saps energy, lowers resistance to infection, prevents the full development of potential intelligence, and shortens life. We have enough food to maintain *today's* world population at *that* level, but no more.

Against all the gloomy prospects of too many people and too few resources, including food, it can be argued that the situation is not really new. Those who refuse to accept prophecies of doom point out that the developed nations, at least, have seen it all before. They have lived through a time when the great majority of people endured slow starvation. Their populations, too, swelled at unprecedented rates during the years of the Industrial Revolution, although not as fast as the present rate in the poorest parts of the world. A century ago, however, say the optimists, the techniques for coping with vast populations and rapid increases were still being worked out, whereas today we have all the techniques we need. As for resources, they were even then considered woefully inadequate for the industrialized countries' foreseeable demands, and some Cassandra or other has again and again prophesied an imminent exhaustion or depletion of this or that resource. Always, however, we have discovered vaster reserves either in the nick of time or well before it.

Who, 15 years ago, would have predicted such vast oil deposits as have been found in the North Sea? Who, even now that the North Sea wells are in production, can speak with any confidence of

Total average calories per head per day

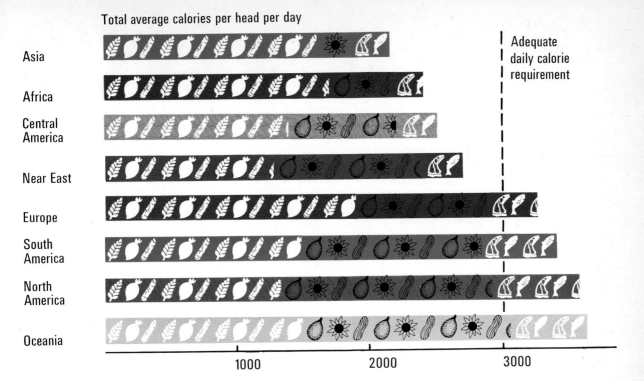

Asia

Africa

Central America

Near East

Europe

South America

North America

Oceania

Adequate daily calorie requirement

1000   2000   3000

The chart above shows the average daily calorie intake in nine major world regions. In four of them it is less than adequate. The chart on the right shows how the world's available food is distributed. In three regions, Europe, North America, and Oceania (Australia, New Zealand, and some nearby islands), there is an excess—the colors reach beyond the radiating lines. Even if all the food were distributed evenly, though, the colors would reach no farther than the broken circle. Thus, the whole world would undergo moderate starvation.

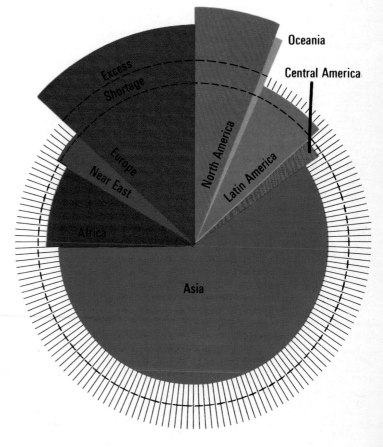

Oceania

Central America

Excess

Shortage

Europe

Near East

North America

Latin America

Africa

Asia

Left: when pictures showing starvation conditions in the Warsaw Ghetto under Nazi rule were published at the end of World War II they shocked the world. Today starvation is just as acute in many of the world's underdeveloped countries, and pictures of equal horror are too commonplace to create the same sort of stir.

their true extent? Who can say what lies farther out on the continental shelf, deeper into the oceans? How much oil may not lie over the great salt domes off the coast of western Africa? And what of the other riches of the ocean? For years now we have known that rich deposits of manganese lie on the deep-ocean floor, and we have barely begun to explore the possibilities of tapping the other resources of that 70 per cent of the earth's surface. The nodules on the deep-ocean floor also contain cobalt, copper, nickel, and various other metals. Several million dollars have already been committed to nodule recovery, and one specter that haunted a recent ocean-mining conference was the danger of *depressing* the world price of minerals.

As traditional resources run out, the optimists argue, it will surely become economically and technically feasible to find and tap as-yet-undiscovered resources; we can direct our surplus wealth to those ends, just as we have recently been diverting it into oil because of the scarcity of that commodity. No one denies that the problems are unprecedented in scale and in the speed with which they are multiplying, but our technological skills are also unprecedented. If we fail, say the optimists, it will not be because the problems were insuperable, nor because our technology was inadequate, but because we lacked the will and the wisdom: we made the job impossible by believing those pessimists who "proved" it could *not* be done.

It is possible to appreciate both sides of this debate even if you agree with neither. One side says we cannot go on as we are; the other says we can—for a while, anyway—and there is no point in trying to guess our grandchildren's needs and aims. Perhaps, though, there is a middle ground. At any rate, recent years have seen an increasing number of books, papers, TV programs, and lectures in which people argue an alternative to this can/cannot battle. Some of the middle-grounders point out that the two sides have more in common than they may think. Very few of the optimists would maintain that population can increase indefinitely. However low the rate of increase, there must be *some* limit at which no extra mouth can be fed. Everyone must agree with that statement. Most people must agree, too, that our present technology is wasteful of both resources and energy, and that a major result of the waste is pollution, which fills the environment with undesirable substances that the environment is not capable of absorbing, breaking down, or neutralizing.

Another point of agreement is that our swelling populations are increasingly hungry for land. Sooner rather than later, the earth's wilderness will vanish forever—or, at best, those that survive will exist only in designated, legally preserved areas. Down at the tail end of this chain of reasoning are the plants and animals with which we share the earth, and which are utterly at the mercy of our activity, or lack of it. This tail has a sting in it, for plants and animals are not naturally organized like a vast outdoor zoo,

Among most of the world's poor, a pot belly is a sign of severe malnutrition (left). But among the world's more privileged people (right) it is the result of appetite and abundance combined.

Drawing by Chas. Addams; © 1969 The New Yorker Magazine Inc.

Now maybe they'll be moved to do something about water pollution!

awaiting man's pleasure. They live in balanced, self-regulating ecosystems, most of which are fairly adaptable and resilient, but all of which are capable of being put wildly out of balance. A species such as man, who even at the best of times has only four to six weeks' supply of food in store, would be very foolish to keep pushing those ecosystems—the ultimate source of all our nourishment—toward the point of imbalance.

For all these reasons, then, many thoughtful people, rejecting both total pessimism and do-nothing optimism, believe that our best hope lies in finding an alternative to the "world doom" of the recent past. They contend that self-interest as well as altruism should force us to look at our goals afresh—to ask whether domed megalopolises and wall-to-wall TV *are* still our dream. If not, the time to say so is now, before the poorer countries follow the rich ones into this dead end.

The alternatives proposed are as varied as any student of human nature would predict. Among them you can find at least one to correspond with every type of society mentioned in the Aristotelian theory of political cycles that we discussed in the first chapter of this book. Yet, for all their variety, they have many common factors and views in addition to their unanimous rejection of the megadream. In particular, as anyone who has studied historical visions of utopia will notice, there is a refreshing realism about today's proposed alternatives. Traditional utopias have been peopled almost exclusively by a mildly philosophical white bourgeoisie, drifting through Arcadia in plastic druid gowns. No one ever gets mugged or raped. No homes are burgled. "Manufactured" is a synonym for "built to last," and "politician" is another way of saying "good and wise lawgiver." One wonders how grown men could take such visions seriously, in the light of 10,000 years of human civilization!

Today's typical blueprint for a better society makes for poor rhetoric (which may be a mark in

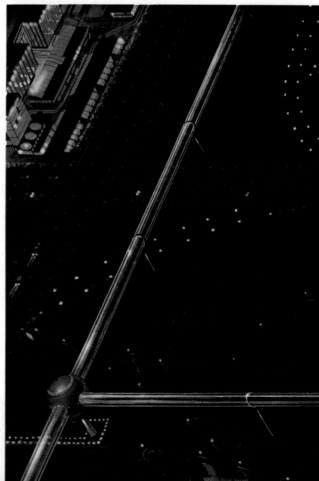

Visions of utopia are often more characteristic of the age that creates them than of any actual future. For instance, Thomas More's Utopia (the book that gave the word to the language), written in 1516, pictured utopia entirely in late medieval terms, as is shown on the title-page illustration (upper right). Similarly 20th-century utopia (right) is soundly based in present-day and foreseeable technology—on seeing it you almost ask "where is that?" instead of "what is that?" An 11-year-old boy of the late 20th century has painted his vision of utopia (above) that is less dominated by technology. Although the "ripple-carpet" transit system is beyond today's capabilities, it is balanced by a liberal provision for art and recreation. A portent, perhaps, of life in a post-industrial world?

its favor), because it specifies, with clearsighted realism, not that there should be *no* crime, but that murder should be kept down to an annual rate of about 2 per 100,000, rape to about 10, and assaults to about 20. Similarly, the modern visionary is likely to express a hope that perhaps as much as one quarter of all corruption will be discovered, exposed, and punished, and so on.

Most proposals foresee a widespread decentralization, not just of government but of industry and technology too. This should be brought about in part by the development of small, cheap, nonpolluting local power sources, such as various devices for trapping solar power. Some of these devices are expected to produce the kind of low-grade background heat for which we now burn high-grade fuels, and others will convert sunshine directly into electrical or motive power. There are also suggestions for the indirect use of solar energy: for example, alcohol (for fuel) and other complex chemicals can be extracted from corn and other plants. Some forward-looking thinkers see wind and water, too, as potentially valuable sources of local, small-scale energy. Some of the many proposals for substitutes to the fossil fuels are worth looking at in depth.

Apart from anything else, they show that "low-impact technology" is by no means merely a revamped technology from horse-and-buggy days.

The obvious way to trap the sun's rays without disrupting nature is to use the method that nature herself evolved some 2000 million years ago: photosynthesis, the process by which the green pigment of a plant catches solar energy that permits the plant to change atmospheric carbon dioxide (CO_2) into sugar. So would it not be feasible to get great quantities of solar energy indirectly by extracting it from the green plants? Well, the trouble is that most plants trap less than a fraction of 1 per cent of the total solar energy that falls on their leaves. What limits them is the availability of CO_2, which normally comprises no more than about 0.03 per cent of the atmosphere. No matter how much water and other nutrients there are in the soil or how smiling the sun, the low availability of CO_2 limits the efficiency of photosynthetic conversion. Thus, it would take unimaginable amounts of most crops to give quite small supplies of power.

One man who has investigated the problem is biologist James Bassham, of the Lawrence Berkeley laboratory in California. He points out

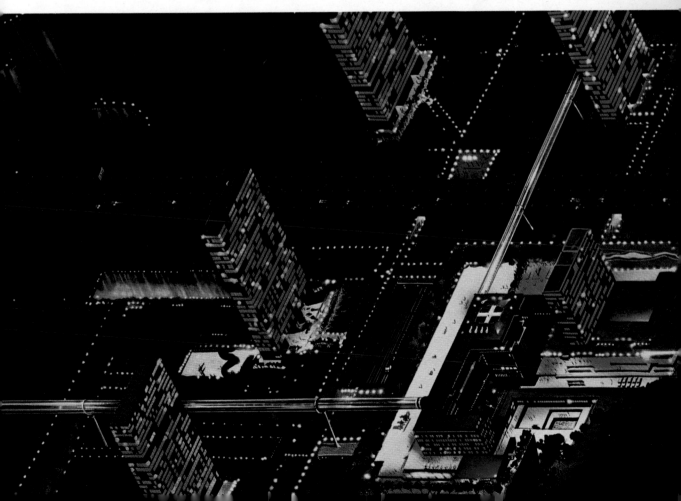

that some plants, such as sorghum and sugar cane, have their own internal chemical pumps, which steadily maintain a high concentration of CO_2 around their green pigment; as a result, they can convert solar energy as much as 10 times more efficiently than do other plants. To grow such plants as energy crops rather than as food, however, would require the use of valuable agricultural land, which is already scarce. An alternative would be to plant a crop such as alfalfa in marginally fertile or even desert land where it could be grown under plastic sheeting in an atmosphere where CO_2 content would be artificially enriched to around 0.2 per cent. This would raise the conversion efficiency to at least half that of sugar cane at its best—and it would yield 10 tons of protein per acre per year as a by-product. The protein would be extracted first, and the residue digested to yield methane.

Another biologist, Clinton Kemp of Inter-Technology Corp, has also looked at the possibility of using green plants for power. Attacking the problem from the other end, he has suggested that the first thing to do is to specify the require-

Two pictures that contrast two approaches to the harnessing of solar energy. The "low-impact" technology approach is to trap the sun's energy in water-filled panels in the garden and store the heat in drums inside, as in this New Mexico home.

ments of a new energy source, and *then* to look for something in nature that fits those requirements. These are his specifications: the new source should use marginal or unfavorable land, should pose no potential hazard, should be ecologically inoffensive, should rely on existing technologies, should be capable of large-scale application in the late 1970s or 1980s, and should be forever renewable. The answer, as he sees it, is the "energy plantation." After studying a wide variety of existing crops he has found, for instance, that corn could yield energy at competitive prices with coal and oil, but that it would be unsuitable in other ways for an all-year-round source of energy. Like many other crops, it can be harvested at only one season, and so there would be difficult storage problems. Instead of such obvious crops as alfalfa or corn, Kemp and his co-workers at Inter-Technology have been experimenting with deciduous broadleaf trees, especially hybrid poplars that resprout from the stump after cutting. At 3700 trees per acre, the energy yield would be about 120 million Btu on marginal land. (One Btu, or British Thermal Unit

The "high-technology" approach is used at the solar farm high in the French Pyrenees where giant panels (not visible here) track the sun and beam its rays onto a vast focusing mirror. Temperatures up to 6300°F have been achieved.

will raise the temperature of a pound of water by one degree Farenheit.) Inter-Technology estimates that around 500 million acres of such land (which is of little use for growing food crops) are available in the United States alone. If planted with poplars, those acres could yield energy equivalent to all the generating capacity currently installed in the country.

Once you had harvested the sugar or corn or alfalfa or poplar, how would you tap the energy it has stored? The system most favored would be to "digest" it in air-free tanks something like giant septic tanks. Microorganisms within them would convert the cellulose and carbohydrates in the crop to methane, the chief constituent of natural gas, and after digestion, the residue in the tanks would make an excellent fertilizer. When methane is burned, it yields water and CO_2. So there would be no toxic by-product and no pollutant in the whole cycle.

The US Army's Natick laboratories are working on a related, but somewhat different system of extracting energy from plants. Curiously enough, their work has developed from studies of how to stop microbes from digesting the cellulose in uniforms and equipment. In the course of their efforts to produce microbe-resistant materials, army scientists have identified over 13,000 microorganisms that naturally break down cellulose. The worst offender is a type of fungus, *Trichoderma viride*, that was first identified during World War II, when it was discovered feasting on a discarded ammunition belt in the New Guinea jungle. This tiny organism can break down even the most crystalline and indigestible forms of cellulose, and the Natick biologists have bred strains at least four times as efficient as the wild strain. The product of digestion is not methane but a crude glucose syrup, which can be used as the feedstock of chemical processes. Among its end products could be yeast-type cells that would yield protein and alcohol, ethanol (an alternative to gasoline), or industrial solvents.

We need not depend on land plants alone as an indirect source of solar energy. If we turn to the oceans, the possible yield of energy from photosynthesis becomes mind-boggling. One pioneer experiment, by the US Navy's Undersea Center, is already under way at a seven-acre seaweed farm near San Clemente Island off the coast of California. The seaweed is kelp, one of the fastest growing of all plants. Navy frogmen set out roots of kelp on rafts that are kept down 40 feet under the surface by being tethered to the bottom. The cold water near the bottom, which lies 260 feet below the rafts, is rich in plant nutrients; and when it is brought toward the surface by artificially created upwellings, the kelp thrives. Low in cellulose but full of other organic compounds, it could prove useful either as a food or as the source of a whole spectrum of chemicals that we now process from coal or petroleum—waxes, fertilizers, solvents, dyes, lubricants, fuels, and raw materials for plastics and man-made fibers.

The experimenters have suggested that a square mile of kelp farm might be made to feed at least 3000 people *and* yield an energy by-product big enough to support 300 of them at current US energy-consumption levels. At the very least, 80 million square miles of the oceans— about 57 per cent of the total water surface—are capable of being farmed in this way. Thus, if techniques for full-scale kelp farming were perfected, the oceans could support around 24,000 million people (six times the present world population) at the equivalent of America's high living standards of the late 1970s. All this would be achieved without farming a single acre of land, pumping a single barrel of oil, or building one more nuclear power plant!

Of course, it should not be inferred that we can therefore justifiably aim at reaching such astronomical population levels. But the figures do show how great and untapped is the sun as a potential energy source, and how we *could* tap it without developing any startling new technology, yet without condemning ourselves to a rugged, pioneering-type existence. There is even the possibility of tapping the electric energy of photosynthesis directly, either in the form of electric current or as hydrogen gas. Donald Krampitz at Case Western Reserve University in Ohio has achieved the production of hydrogen by this route in the laboratory, and he believes that a full-scale system could have an efficiency rate of up to 10 per cent. Since plant photosynthesis is generally only a fraction of 1 per cent efficient, such an informed suggestion that we can learn how to capture as much as one tenth of the sun's energy raises staggering expectations.

What is so promising about the work of Krampitz and the others is that the product they are trying to manufacture is not just energy but an actual fuel. People often argue against solar energy on the grounds that solar-cell technology is costly and works only when the sun shines.

Right: our likeliest route to the efficient capture of solar energy is via plants, which do it all the time. John Wyndham's apocalyptic vision in Day of the Triffids *may provide an allegory of our future if we come to depend on energy plants as we now depend on oil. Triffids were mythical plants cultivated for their rich vegetable-oil yield. But they could move and had lethal stings. In Wyndham's story the triffids got out of hand and almost destroyed civilization.*

Below: at a practical level one possible candidate for harnessing solar energy is the sugar cane, which has special systems for concentrating CO_2 within itself. The low natural concentration of CO_2 limits conversion efficiency in most other plants. Already it is possible to make gasoline and other industrial chemicals from sugar or molasses.

Then, too, a huge loss of its product, electricity, is caused by having to store it up in electric cells. "You can't run a car on solar power" is how the argument is usually summed up. What we are really short of, in fact, is not energy but a clean fuel, free of sulfur and nitrogen compounds. The path to a solution of these problems is likely to be found along the biotechnology route that I have been mapping out. Moreover, the use of plants for trapping energy is an answer not only to the arguments against trying to utilize solar power directly, but also to objections to nuclear power.

A study made as recently as 1975 indicates that to bring the anticipated world population of the year 2075 (which appears likely to quadruple our present population) up to existing levels of energy consumption in the industrialized nations would call for 3000 nuclear "parks," each with eight fast breeder reactors, each such reactor being about five times bigger than today's biggest. Such a program would require the building of four reactors every week from now until the year 2000. Thereafter, to replace outworn reactors as well as add new ones, the program would need to be stepped up so as to produce two reactors a day. Even knowing how

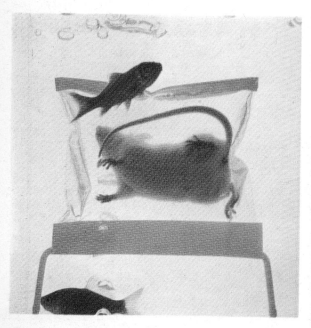

Right: an artist's impression of a deep-sea city. The technology for exploiting the ocean depths, either remotely with machines or in person, is already well advanced. For instance, the man-made rubber known as polydimethylsiloxane allows dissolved gases to pass through it either way with great efficiency—so much so that a rat sealed in a bag made of it can breathe indefinitely underwater, as shown above. Fresh oxygen passes in through the rubber to replace expired CO_2, which passes out.

foolhardy it is to maintain that any further development is "impossible," we must still admit that the strain of such an ambitious building program—not to mention its terrible dangers—would make *any* attempt at decentralization of energy resources (such as biotechnology might achieve) worthwhile.

Decentralization may well be our goal not only in the field of energy but in other fields, including government. Let me emphasize, though, that decentralization is not the same as dispersal. Many futurologists foresee a world in which big cities still exist on more or less their present scale—unless, for some reason, world population falls to a fraction of its present level. They believe, however, that governmental power and decision-making will (or should) be returned to smaller groups and communities, even if this might mean less efficiency, more collusive corruption, and a great diversity of standards. People, say some of our modern prophets, need to be encouraged to share local problems, local activities, and local decisions, where they can have some influence. Nowadays, when the focus is national or international, an individual may feel powerless and frustrated; this tends to

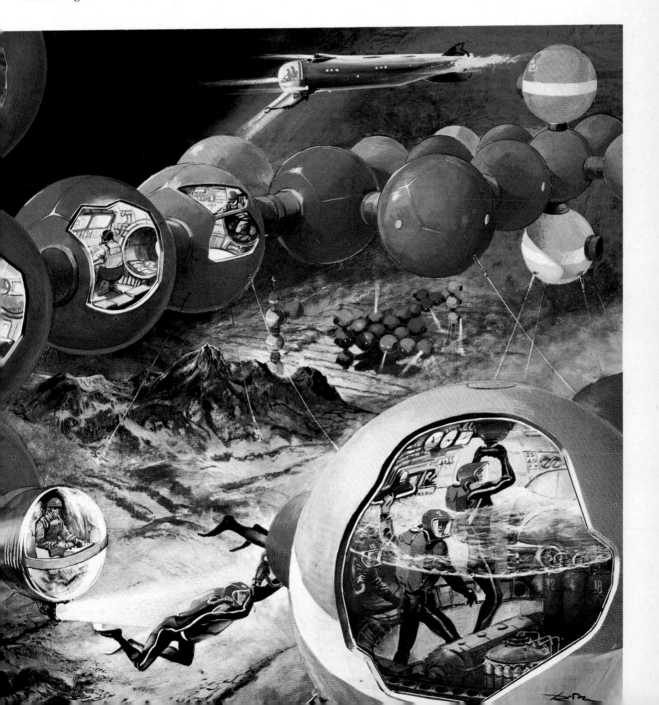

alienate him from modern society, which he sees as a juggernaut, and so to drive him increasingly inward to a private-consumer world or to private fantasies of revolution.

There are good reasons for arguing that decentralization would be a worthwhile development even if it meant an end to the big-city configurations that are so much a mark of our time. Many urban studies have shown that increasing city size does not necessarily go hand in hand with increasing efficiency and economy—that, indeed, it is likely to cost more than twice as much per citizen to run a city of 700,000 as to run one of 100,000. It may or may not be true that bigger cities can deploy their resources better and can practice certain economies that smaller organizations cannot achieve, but any such economy is more than offset by costs that do not bedevil smaller cities. One major cost is that of crime. The figures quoted a few paragraphs back

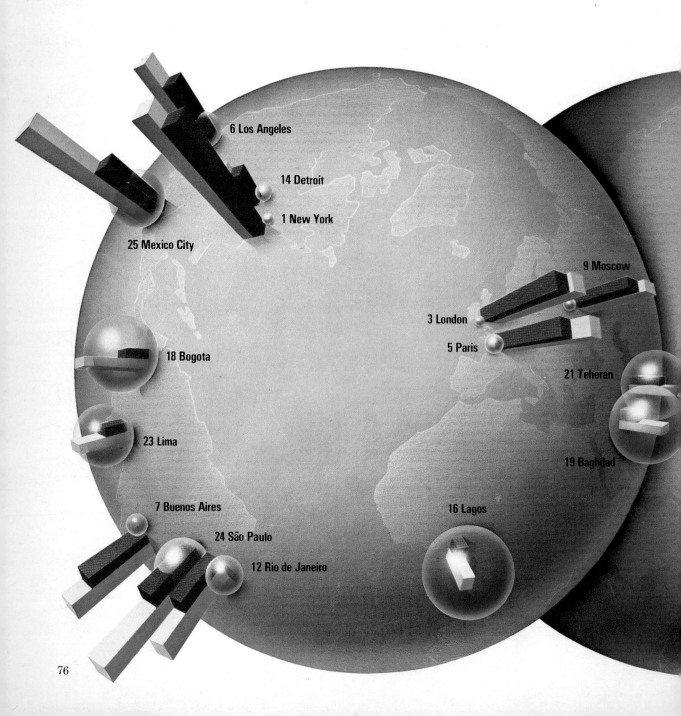

as "ideal" crime rates for a better society in the future—2 per 100,000 for murder, 10 per 100,000 for rape, etc.—are actually fairly typical of present-day American communities of under 10,000, with a population density of about 1700 per acre. If we turn to the statistics for cities of 250,000, with densities of over 7000 to the acre, we note a threefold rise in the murder rate, a fivefold increase in rape and other kinds of assault, and a tenfold leap in robbing. There is a further leap from here to the cities with a population of half a million, and so on.

A development of local energy sources and a decentralization of power within city limits would both be partial reversals of the trend toward "bigger and better" that has accompanied industrialization during the past two centuries. So, too, would be another feature common to most plans for an alternative to our present way of life: an increase in self-sufficiency. In fact, this

The World's Crisis Points

The biggest cities in the world (1-14) have either already reached population saturation point or face heavy further population influx in the next decade. The populations of the fastest-growing cities (15-26) will all have doubled by 1980.

Biggest cities (population in millions)	1970	1985	Growth rate (%)
1 New York	16.3	18.8	15
2 Tokyo	14.9	25.2	69
3 London	10.5	11.1	6
4 Shanghai	10.0	14.3	43
5 Paris	8.4	10.9	30
6 Los Angeles	8.4	13.7	63
7 Buenos Aires	8.4	11.7	39
8 Osaka	7.6	11.8	55
9 Moscow	7.1	8.0	13
10 Peking	7.0	12.0	71
11 Calcutta	6.9	12.1	75
12 Rio de Janeiro	6.8	11.4	68
13 Jakarta	4.0	7.7	93
14 Detroit	4.0	4.9	23
Fastest growing cities (population in millions)	1970	1985	Growth rate (%)
15 Bandung	1.2	4.1	242
16 Lagos	1.4	4.0	186
17 Karachi	3.5	9.2	163
18 Bogota	2.6	6.4	146
19 Baghdad	2.0	4.9	145
20 Bangkok	3.0	7.1	137
21 Teheran	3.4	7.9	132
22 Seoul	4.6	10.3	124
23 Lima	2.8	6.2	121
24 São Paulo	7.8	16.8	115
25 Mexico City	8.4	17.9	113
26 Bombay	5.8	12.1	109

2 Tokyo
8 Osaka
22 Seoul
10 Peking
4 Shanghai
20 Bangkok
11 Calcutta
15 Bandung
17 Karachi
26 Bombay
13 Jakarta

Projected population 1985
Actual population 1970
Percentage growth rate

Critics of industrial civilization often focus on the city as its most dehumanizing agent. It is too big, they say. People feel no sense of belonging. Society has become a faceless and heartless bureaucracy. Only violent protest, it seems, will guarantee a hearing. In many countries impromtu street violence has escalated into organized urban guerilla warfare. Clashes like this confrontation between police and demonstrators which took place in Amsterdam in 1975, have occurred in most of the world's large cities.

would reverse a trend that began with the earliest city civilizations of man's history.

Total self-sufficiency is an unreachable goal for most people, of course—at least, for those who want to live in fair comfort and sanitary conditions, according to acceptable Western standards. Those who in our own day have achieved a good measure of self-sufficiency have found the task to be a full-time occupation that absorbs almost all their physical and mental energy. Two such individuals are John and Sally Seymour, an English writer and his artist wife, who have lived in self-sufficiency (or close to it) for about 20 years. In 1973 they published a book,

stopped somewhere short of assembly-line ultra-specialization. Groups of interdependent families are only one of many possible ways of achieving collective self-sufficiency. Among others are the kibbutzim of modern Israel and the new-style communes that have such a strong appeal for today's disaffected young people.

Although complete self-sufficiency, whether independent or interdependent, is an unattainable goal for most individuals as well as for mankind, its advocates insist that there is plenty the individual can do to go part of the way. An essential beginning is to pause before each action, each purchase, each consumption, each trip, each jettisoning of garbage, and ask oneself, "Can I do something here and now to reduce the impact of this activity on my environment?" The effect of answering that question by actually doing something is a step toward self-sufficiency, a step away from reliance on large-scale industrial technology. Composted garbage creates fertilizer, which saves the fuel and chemicals that would have gone to make artificial fertilizer . . . and so on. There is quite a lot that you can do short of creating your own private power supply by fixing windmills onto your roof.

There are three basic sets of arguments generally brought forth by those who advocate such a course for anyone who cares about tomorrow's world: moral, environmental, and pragmatic. The moralist claims that the naked pursuit of acquisitions and material affluence has been disastrous. It has encouraged greed and anti-social behavior at every level; it has led to alienation, disaffection, and violent reaction; worst of all, it has not even delivered the contentment it once promised. In short, materialism has failed us, both directly and indirectly. The environmentalist points to the increasing scope and violence of the damage we have been doing to both the living and non-living worlds, and he warns that we are running out of space for further pollution and destruction. Finally, the pragmatist draws on both arguments, and adds that, because time is running out on us, we had better learn to live in harmony with the world and one another for the sake of our own survival, regardless of whether it would also be morally and environmentally right.

Thus, there is a subtle but important difference between the moralist and the other two proponents of a shift away from technological extravagance. The environmentalist and the pragmatist

called — not surprisingly — *Self-sufficiency,* in which they pointed out, among other things, that their chosen way of life became easier for them whenever a like-minded person came to live in their neighborhood. They could then trade, for example, mutton for salt fish, milk for bacon, hand-made pottery for wooden vessels, and so on. John Seymour believes that a group of half a dozen families, each with some training in preindustrial farming and one specific type of handicraft, could make a very good life for themselves. Such minor specialization would not necessarily be the first step on the road toward Detroit. That road, Seymour argues, should have

After a century or more spent in the pursuit of material success, the peoples of the industrial West have recently undergone or witnessed a profound change in orientation. It may be the beginnings of a new, post-industrial civilization or it may be just one more passing fad. It has shown itself in a widespread movement toward communal living, in which the traditional two-parent family is enlarged to a group of like-minded friends. Often they take pride in building usefully from what industrial civilization has discarded as useless, as in the building complex above. Also they try to use the free energy around them in the sun and the wind (lower right). The influence on "straight" civilization has been enormous, even in small things, such as the foam-plastic version of the old haybox, no-fuel cooker (below).

do not necessarily want to unscramble the whole of modern technology, whereas the moralist tends to preach a return to a primitive existence in small communities. To the others this seems like throwing out the carpet along with the dust. Technology is not all bad, by any means. Would it really make sense to scrap a modern pesticide that has a highly selective action and disintegrates quickly—and thus to encourage plagues of locusts? Does anyone really want to scrap our antibiotic technology to the point where children would once again be permitted to die from meningitis or septicemia?

And so the focus of argument necessarily shifts from specifics to a general critique of society. Granted that we cannot contemplate a return to the horse and buggy, what kind of post-industrial society can we realistically envisage? Again, there are dozens of possible substitutes for what we now have, and some futurologists have tried to describe the societies they foresee in great detail, right down to their legal systems. In broad terms, though, all the authorities visualize a world in which there will be no further overall growth, whether of population, of economies, of resource depletion, or of pollution. Perhaps the trend may even be reversed where population and pollution are concerned. Growth will still be possible in some sectors, and so will "progress" in the present material sense; but most forecasters expect such progress to be in communal rather than private services—public

standing of the world and our place in it. Even high technology in the form of computer development and progress in nuclear energy, for instance, goes on. But none of it leads automatically to a higher material standard of living. Life in general may be considerably enriched, but not in terms of individual acquisitions.

At the personal level, the citizens of the early 21st century have found some other way of measuring status than by income and material goods. They may have electronic gadgets on their persons and in their homes that we would envy— the instant communicators and computer access that Philip Handler visualized in his megadream —but they use them for many of the practical purposes for which we use cars. The focus of their world is much more local than ours, and a lot of their leisure time is actually community time. They shop almost entirely locally, work locally— possibly even at home—entertain locally, and concern themselves to a surprising degree (by our standards) with local affairs. Even if they work for a multinational company, their branch of the company has a great deal of local autonomy. In everyday affairs they are much more self-reliant than our generation; they have a degree of self-sufficiency that absorbs a large slice of their leisure hours. For instance, they make things that we buy, repair things that we junk, grow things that we reach for in cans or freezer packs. And, perhaps, they accept for real the home-spun, patched, faded, and frayed life style that it was fashionable for the youthful rich to flirt with back in the 1970s.

To many people alive in the 1970s, the post-industrial world would seem dull indeed. These homespun activities offer few familiar satisfactions and very little on which to shape a dream. No one asks what's your hurry, for there is hardly anywhere to hurry *to*. People probably rant against inequalities—they no doubt always will—but by the standards of the 20th century the inequalities of the 21st are small.

Another thing that many of us would dislike is the quite extraordinary amount of discussion, consultation, buttonholing, and mutual back-scratching that has to be got through before any

transportation rather than private, for instance, or new pharmaceuticals rather than suntan lotions. For a product or a service or an institution to exist, it will no longer be enough to prove mere demand for it. The criterion will be, not "I want," but "I need, and I can prove that need." And the proof will have to involve the social and environmental costs of what is proposed.

From all the foregoing we can try to paint a general picture of post-industrial society as it might emerge in accordance with the most pragmatic of dreams. It is because industrial growth has stopped that we call it *post*-industrial. Science goes on, however. Invention goes on. The study of man, of medicine—all these activities continue to broaden and deepen our under-

undertaking can get started. There have to be curbs on the aggressive, acquisitive impulses that mankind is heir to—and curbs mean restrictive laws. Thus, the number of laws and regulations that frustrate all change, novelty, and "progress" (in the old, 20th-century sense of the word) might seem intolerable to the average man of the 1970s—even to those who used to yearn for stability and complain of the rat race, of the bewildering pace of change, of the noise and bustle, the overcrowding and sheer beastliness, of modern urban life.

Post-industrial society has other disconcerting features. For instance, a vast amount of social manipulation goes on, quite nakedly. People laugh at it, scorn it, and say it doesn't affect them. Yet they tolerate and even connive at it. Curiously enough, they let themselves be manipulated, in a grudging, half-willing, half-reluctant fashion. In our own time we have something like it in the disguised-yet-open publicity that goes on all around us: people on chat shows who just happen to have a book/film/play/novelty coming out shortly . . . special-feature pages on home-freezing/judo/scuba diving, etc., whose "editorial" matter just happened to mention every advertiser who buys space in the paper . . . the politician who found a new angle on which to pin some forthcoming legislation or proposal. . . . We are even manipulated in a different way by undisguised, legitimate bought-and-paid-for advertising, whose total effort is to make us think we *need* all the things the advertisers want to sell. Post-industrial persuasion is like that, except that the product it "sells" is no less than a balanced, static, orderly society. Such persuasion is not called advertising, of course; it is called "psychological engineering."

To us go-getting children of capitalism there is something vaguely and unpalatably communistic about all this. Nor does it help to know that for 20th-century communists our vision of the post-industrial world is hateful. The individualism, self-reliance, and local autonomy of the new society, along with its nondialetical, multivalued approach to day-to-day problems—these have nothing to do with the good life according to the devotees of Marx and Lenin.

Transportation problems have brought many industrial cities to their limits. New York's crowded streets (below) offer a scene that is familiar around the world. Japan's subway employs special "packers" to cram people into trains (right). In the post-industrial world, telecommunications will substitute for much of this physical transit.

The post-industrial people themselves see their society from a somewhat different standpoint. Mainly they are glad to have escaped the horrors we predicted for them, horrors that, by our feverish struggle for growth, we almost bequeathed them. They do not like us much. In fact, they think of us as selfish, arrogant, often childish, and always short-sighted. They read our predictions of doom, depressing novels, and despairing poetry; they listen to our songs of protest and watch old movies of our riots, burnings, small wars, and massacres; they rerun our documentaries on poverty, starvation, and the less edifying aspects of affluence; and they shudder with relief. Among them are a few old-timers—really old-timers, survivors of our days—who try to tell post-industrial men and women that things were not really that bad back in the 20th century, at least not for most people. Their words, though, are no more believed than we heed the words of our remaining Victorians, who tell us the same thing about their century.

Post-industrial people reckon that it is worth a lot of sacrifices to have avoided the life their great-grandfathers—we—endured and the even worse fate we were stoking up for them. Naturally, there are plenty of doomsayers still around, warning that mankind is not yet out of the woods. The shade of Malthus still stretches far into the future. We moved mankind close to the brink of catastropic collapse; we halted the process just in time. But our great-grandchildren will not yet have moved far enough back into the realm of safety to forgive us. Their margin for maneuver will still be desperately narrow. Like all survivors after a close shave, they will be glad to stay in line and keep a low profile.

In any case, as they will rightly say, you can hardly call post-industrial life dull in the sense that everything is uniform. Indeed, local autonomy will ensure the very opposite. Anybody who dislikes his community will be free to leave and find another one to suit his tastes, "In the *industrial* age," they sneer, "it was getting so you couldn't tell Birmingham from Bangkok." Post-industrial differences, though, will be like those between neighboring tribes, and not like those between megalopolis and the backwoods.

One thing will puzzle some of the new generation: the speed at which it all happened. One minute the world appeared to be hell-bent on growth; the next minute, seemingly, everything had become post-industrial. Without a revolution, with very little fanfare, with just a minor shuffle of government . . . it is hard to put your finger on exactly where and when the turnabout came. But the old-timers among the post-industrialists will say that it did not happen all that abruptly. They will point out that our age,

Past utopians naïvely expected a future without violence. To-day's visions allow for vicarious outlets for hostility, as in writer William Harrison's 21st-century game Rollerball—"the most vicious and brutal game ever." Watched by billion-strong audiences, the game substitutes for war in a war-less world.

the industrial age, was actually littered with signposts to theirs. And so it is. To return from the dream to present-day reality, we have our ecologists, Friends of the Earth, zero-growers, self-supporters, consumerists, vegetarians, health feeders, doomwatchers, proponents of labor participation in industrial management, power sharers, recyclers, Peace-Corps people, internationalists, community-service volunteers, and so on. It is perhaps true that a majority of late-industrial thinkers will soon be devoting part of their mental and physical energy to the themes and pursuits that will shape the world of post-industrial man. Maybe it would not be stretching truth too far to say that the foundations of post-industrial society are already being laid here and now in the 1970s.

The chief beneficiaries of the anticipated turnabout in human values will not be the human species, however, for man is ruthless and intelligent enough to survive almost any man-made catastrophe. The greatest gains will go to the rest of the living earth—to all the animals and plants whose survival we put in jeopardy.

The Future of Life

There is a saying to the effect that no matter how many good manners you try to teach a child, it ends up behaving like its parents. If the future is the child of the present, a similar rule seems to operate; despite thousands of years of moralizing, of planning, of utopian visions and revolutionary fervor, tomorrows and yesterdays have always had more in common than the proponents of change like to acknowledge. Usually, then, today is a good guide to tomorrow, just as the clouds or the blue sky on the windward horizon are a good guide to the coming weather. So what sort of parent is today? What sort of future does it promise for the earth in general and for the life that earth supports?

Unfortunately, most of today's offspring seem to be little horrors, all set to turn into big horrors. The post-industrial scenario with which we ended the previous chapter may turn out to be ridiculously optimistic. The present holds nightmares that are still able to wake us up bathed in sweat.

Some of the nightmares are familar ones, embodying mankind's most ancient fears. Our legends are full of tales of universal flood and fire. There is something in most people that is deeply stirred and satisfied at the idea of absolute catastrophe. Watch a child as it lovingly builds a sand castle, and then look at its face as it smashes the construction, and you will see this delight at first hand. As we mature and lose the magical ability that makes a living world out of sand, we demand more "realism" in our fantasies of destruction, and we hide our delight better.

One absolute catastrophe that bridges the ages of myth and science is the notion that the earth

An eruption of glowing helium gas, arcing out over a quarter of a million miles from the face of the sun, is a spectacular reminder of the sun's restless interior and of the instabilities that might provoke it into a runaway nova or supernova explosion.

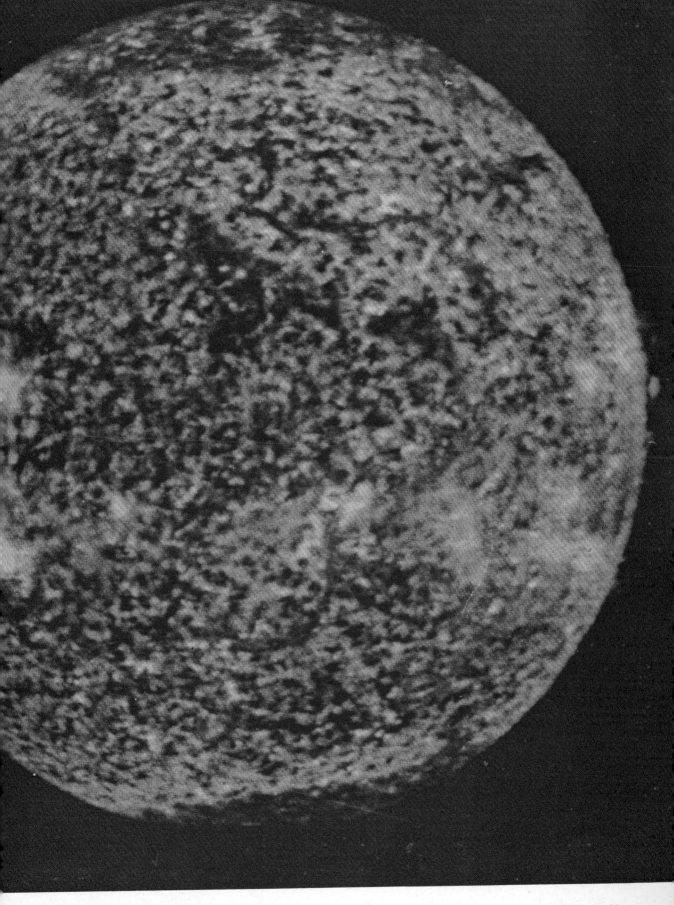

might collide with another celestial body, which is even now on a collision course with us, just a few light-years or parsecs away. An alternative is the idea that the sun will suddenly flare up into a nova or supernova, either of which would bring the earth within the sun's diameter. In the first half of this century, when the nature of the universe was becoming apparent, fantasies of that sort were on every bookstall. They were comforting, in a sense, because there was absolutely nothing that anyone, not even the whole of mankind acting in miraculous concert, could do to avert them, or to prepare to survive them.

Our more recent scenarios lack that sort of comfort. The collapse they envision is man-made. Chief in this group is the possibility of nuclear holocaust involving the two super-powers, America and the Soviet Union, either through escalation of lesser hostilities or by accident. There was a time in 1962—the "Cuba crisis"— when such a possibility seemed both real and imminent. (I remember riding the subway home each day and hoping that the bombs would not fall while I was down there where I would have a real, if slim, chance of survival. Like most of my contemporaries, I had thought a lot about the post-nuclear-holocaust world and I wanted no part of it.) Since then, and maybe because that trauma was so widely shared, the probability of this particular scenario coming true has diminished. Even so, there are plenty of other potential nuclear-war dangers that could lead to something terrible, though perhaps rather less catastrophic than the total collapse of civilization.

Other total-collapse scenarios envisage a new ice age as a result of atmospheric pollution. The countless fine particles we spew into the atmosphere, from jet tailpipes, factories, cars, and home furnaces, plus those that nature puts there, might form a nucleus for cloud droplets to condense around. The clouds might then reflect back into space solar energy that should fall on the earth and heat it. Even the particles that did not form cloud nuclei would play a part in this reradiation. As a result the earth would gradually cool and a new ice age would then begin.

There is an opposite scenario, based on the fact

An exploding atom bomb. Even a "small" Hiroshima-type bomb like this releases enormous destructive energy. The far more powerful hydrogen bomb, capable of obliterating hundreds of square miles, provided scenarios of death on such a massive scale that it haunted mankind throughout most of the 1960s.

Alternative scenarios of apocalyptic doom. The one shown at left foresees the cooling of the Earth as man-made dust and smoke reflect solar heat back into space. The glaciers spread and wipe out civilization. The picture is from a 1929 science fiction magazine cover. The scenario shown at right predicts that increasing CO_2 in the earth's atmosphere will trap more and more solar heat until the ice caps melt, raising the sea level by 400 feet. The diagrams show how London, Paris, and New York would all vanish beneath the waves if that were to happen.

that carbon dioxide in the atmosphere is transparent to the shortwave infrared heat radiation from the sun (or any white-hot body) but is opaque to longwave infrared from warm objects on earth, which means that heat can get in but can't get out so easily. Because the glass in greenhouses acts as a filter in this way, greenhouses are always warmer than the outside air, and so the effect is popularly called the "greenhouse effect." Measurements show a significant increase in atmospheric carbon dioxide this century— 15 per cent or more. Some comes from the fossil fuels we burn so prodigiously, but more emanates

from drying bogs and wetlands, which hold over 350,000 million tons of the gas. Plowing also releases enormous amounts of soil-held gas. Projecting the graph forward, we can see enough increase by 1990 to raise the world temperatures an average 9°F. That much thermal pollution would perhaps be enough to melt Antarctica and raise the sea level by 400 feet. A 20-foot rise would be enough to submerge New York and London.

Meteorologists have calculated the contrasting effects of the two processes, and it seems that the cooling effect could overtake the heating effect during this decade and produce a drop of ·9°F

If the Ice Caps Melted . . .

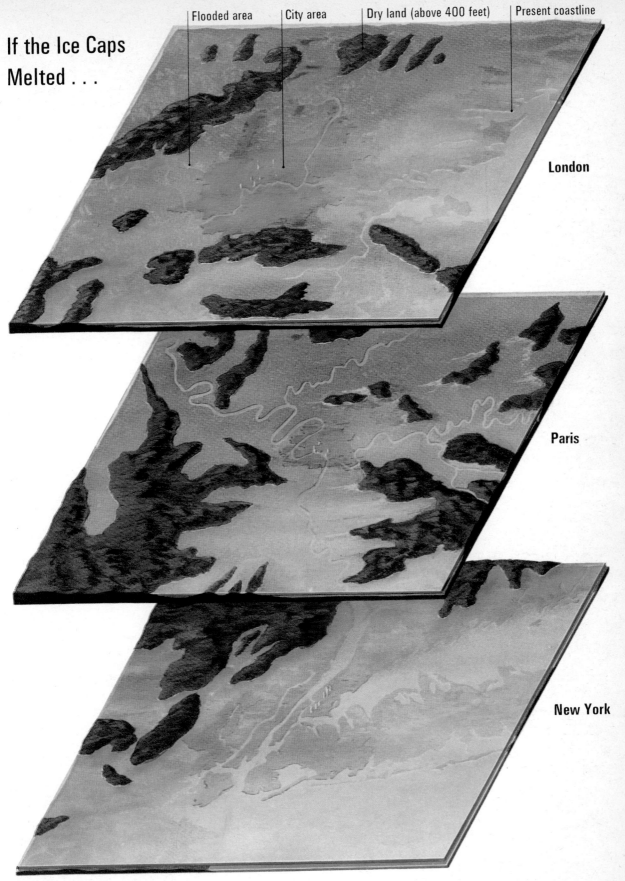

Flooded area | City area | Dry land (above 400 feet) | Present coastline

London

Paris

New York

93

before the year 2000—quite enough to start a new ice age. If Europe and most of North America, together with the equivalent latitudes of the Southern Hemisphere, were so quickly plunged into Siberian conditions, and the tropics became no more than temperate, our present civilization could not survive. Apart from anything else, such a world could feed only about one tenth of its present numbers (some say only one hundredth) at moderate starvation level. Ironically, we should then be looking for all the thermal pollution that we could find. We should not be able to get enough of it.

Thermal pollution is the basis of another collapse scenario. So many factors contribute to the earth's heat balance that it is barely understood. It was only when man-made satellites began to orbit the planet that we started to make the first reliable measurements of heat input and output from different parts of the globe at different seasons and under different conditions. This is far too short a time for us to have noticed any definite trends, much less to have worked out useful equations, and much less still to have framed an acceptable theory about the effect of human thermal pollution by waste heat from industrial processes and forest burning. Estimates of the total effect vary so widely that they are not worth quoting, but we know that it would take only a 10-per-cent increase in the radiation balance to raise terrestrial temperatures by as much as 36°F. At that rate, the poles would become tropical, most fish would die, and the tropics would become too warm for all animals except a few insects. One calculation predicts this fate for sometime in the 1990s.

A related doom might come from the depletion of oxygen. All the free oxygen in the atmosphere comes from plants. It is there only because plants put out slightly more oxygen than the world's oxidizing processes consume. It has been estimated that of every 10,000 units of oxygen thus put out, oxidation consumes about 9999. It is easy to forget the slimness of this margin. It would not take much reduction of the earth's plant and plankton resources nor much increase in our oxygen consumption to reverse the balance. Visions of a complete return to a primeval world without oxygen are probably unrealistic, but even a partial reduction would make our present type of civilization unworkable.

Ecological disasters involving the collapse of

Predictions of this or that catastrophe through our unintentional modification of the weather are based on the long term effects that air pollution (left) might have on the overall heat balance of the earth. Warnings against pollution of the ocean deserve more respect. Its minute life forms such as the krill shown below, are the foundation of a world ecosystem on which all life depends.

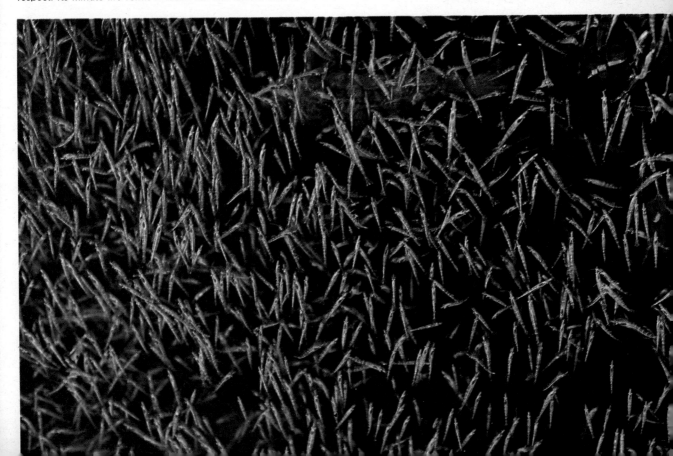

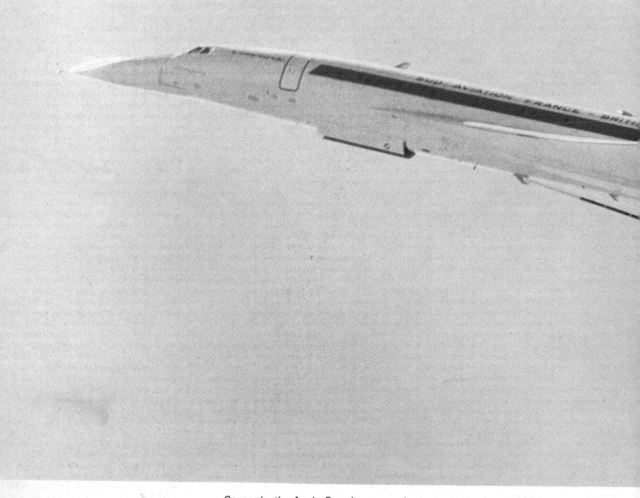

Concorde, the Anglo-French supersonic passenger transport (SST for short) whose noise and exhaust emissions have come in for widespread criticism. Forecasts made in the boom years of the 1960s allowed for 500 SSTs in service by the year 2000.

the world ecosystems could have any of the fore-going scenarios as a secondary effect. One such disaster might be the death of plankton, source of most of our oxygen and foundation of all marine life. This could be brought about by continued dumping of poisons, herbicides, and other pollutants, whether directly into the sea, into the waterways that lead to it, or into the air that covers it. Individual scenarios for other disasters are too varied to summarize, but they all involve one important basic principle:

The physical world and nature have evolved slowly together, each adapting to changes in the other. Even in medieval times, despite the vast deforestation already practiced by man, which had created several deserts, the two were pretty much in harmony. So *anything* we do, *any* change we make, is likely to mean a deterioration in that harmony. This is the real danger in pollution: not that it is unsightly, not that it smells, not that it damages our health, but that it pushes the ecosystem out of balance.

There are other possible horrors in the offing, too. When supersonic transport aircraft (SSTs)

Although SST emissions over 40 years would total no more than the present weekly output of the cars in Los Angeles alone, they would interfere with the stratosphere and dangerously deplete the protective ozone layer.

were first mooted in the 1960s, climatologists pointed to the dangers of liberating so much burned fuel at the 65,000-foot cruising altitude of these aircraft. They warned that the oxides of nitrogen in the exhaust might soak up the ozone at that height. The ozone blanket shields us from radiation; without it we should certainly get an overdose of ultraviolet radiation. Later calculations damped down this fear, but the US Department of Transportation commissioned the Massachusetts Institute of Technology to take a detailed look at the problem. MIT scientists built a

mathematical model of the atmosphere up to 43 miles and analyzed what would happen to its physics and chemistry under the impact of an anticipated 500 SSTs in service by the year 2000. In a report released in 1974, they suggest that the ozone blanket would be depleted by a risky 16 per cent in the Northern Hemisphere and by a lesser but significant amount in the Southern Hemisphere. The resulting increase in ultraviolet penetration, concludes the report, would be one of those massively unbalancing assaults whose far-reaching effects on the existing ecosystem would probably be completely catastrophic.

If all such perils exist now, when only a quarter or a third of the world population lives an industrially developed life, how much greater would they be, how much nearer would they be brought, with every increase in development and every rise in population. Very few thinking people doubt that, in the long run, we shall have to develop a low-impact technology and reduce our numbers. The great question is whether we shall do it sensibly and by cooperative international programs, or be forced to it by a series of close shaves whose effects will be so horrible that the situation may sometimes seem like a total-catastrophe nightmare.

The worst of these "close shaves" would be world starvation, with its traditional attendants: plague and war. This scenario envisages a probable collapse of world civilization and a return, for the most prosperous of us, to the kind of life eked out by the poor in the 17th century. Indeed, that could happen soon and suddenly; after all, we rarely have more than 30 days' reserve supply of cereals anywhere in the world, and sometimes we have no more than 20 days' supply. A likelier prospect, though, is for population slowly but steadily to outstrip food production, as it has already done in many places, and for perpetual undernourishment to deteriorate gradually into outright starvation. In short, there will

Below: the granaries of the Sange Dogon people in Mali, West Africa. Manhandling the grain here is literally man-handling. Obviously, such reserves are available to local consumers only.

be an increasing number of places in the world where the ghost of Malthus could stalk, muttering: "I told you so." The huge predictable surpluses of North America in the 1960s have already vanished. In the 1970s we have seen portents of things to come in the death by starvation of perhaps millions of people in Ethiopia and across the southern fringe of the Sahara to the states of West Africa, and in India and Bangladesh. There will be a great deal more of extreme starvation throughout much of Africa, Central and South

"The Last of England."
By permission of the Birmingham Museum and Art Gallery.

Minamata, a small Japanese seaport, has given its name to a disease (mercury poisoning) and to a new kind of disaster—the unseen, unsuspected poisoning of an area by industry. In 1958 Chisso Plastics quietly stopped dumping toxic mercury wastes into Minamata Bay and turned their waste pipes instead into this tidal pool on the other side of town. When people here developed the symptoms of the strange "disease" that had first appeared in 1953 in the Minamata Bay area, it became clear that the mercury was contaminating the fish that formed a large part of the diet of the coastal population.

America, and Asia. Eventually, too as food exports to Europe and Japan suffer, people in such industrial areas will face a marked deteroration in diet and a reduction in standard of living.

The result, however, is not likely to be total annihilation. It is a fact of history that even the most horrifying starvation levels do not totally destroy a country or its institutions. The Irish potato famine of 1845–8, which by death and enforced emigration actually quartered the population in less than five years, left a country that was scarred, physically and in spirit, but Ireland survived. The economic indicators on the graphs of the period show a dip—but then a recovery. Food exports continued throughout the famine, for the country had to "pay its way" in a world where international aid was unknown and intranational aid was perfunctory. The same harsh economics rule today. Although governments and leaders topple, countries and their businesses limp on, somehow. Moreover, a localized catastrophe may actually have the effect of improving conditions in the wider world.

Critics of computer-forecast futures point out that the computer programs underestimate this last factor. In every country there are haves and have-nots. The haves are not necessarily rich. In fact, everyone may be so desperately poor that the ownership of just a small plot of ground separates the haves from the have-nots. Even so, it is an important dividing line. A great many have-nots can be destroyed without having an appreciable economic effect on the rest of their society, and the effect, when it is finally felt, may be economically favorable (however morally deplorable that fact may be).

By analogy, a similar argument applies to other doomsday nightmares, converting them instead into the close-shave variety (unless, as I pointed out earlier, they bring on a full-scale, global ecological disaster). A nuclear-power-reactor accident . . . limited nuclear war . . . widespread toxic-metal pollution . . . cancer "epidemics" from increasing pollution by unsuspected carcinogens . . . large-scale deaths from organic pollutants that affect the liver, gut, kidney, nerve tissue, and so on—these are unlikely to hit suddenly and on a world-catastrophe scale. It is far more probable that any one of them would cause outbreaks affecting at most, only tens of millions of people. As with starvation, the total effect could be beneficial to the economy in the long term, and it could be more successful than

Gelada baboons (above) and the walia ibex (above right) are two of the unique species that inhabit the Simien Mountains in northern Ethiopia. Human starvation there is so intense that farmers will inevitably move into and destroy these animals' habitat.

any amount of moral persuasion in forcing the survivors to abandon practices or revise procedures that endanger the environment.

The late Dr. Lloyd Berkner, chairman of America's Presidential Scientific Advisory Committee and of the International Scientific Steering Committee, made this point in a 1966 paper called "Truth and Consequences in a New Era." He concluded that people will not change long-held beliefs and deeply ingrained habits unless an irresistible force moves them. "I just fear," he wrote, "that the only force will be [a combination of] hunger, disease, brutality, and death."

One of these forces, hunger, has to some degree already started translating itself into reality. The watershed probably came in a 1974 international conference on food sponsored by the United Nations, when it became grimly clear that aid in the form of food to have-not nations from the developed countries was a thing of the past, and that even economic aid was certain to become less open-handed. Since then, an informal sort of "triage" has been at work. Triage is a system devised by French army doctors for sorting battle casualties into three groups: those who will survive even with no treatment; those whom

Right: many deserts are really fertile land devoid of water. When irrigation is applied, their acres are as productive as regular farmland, as here at Yuma, Arizona. As demand for food rises, areas such as this will increasingly come under the plow.

no treatment, however intensive, could save; and those in between, who can probably survive with help. Because medical aid on the battlefield is limited, the doctors reason, it makes sense to apply it only to the third category. A similar logic now appears to determine the amount and direction of international aid programs. Countries such as Taiwan and Hong Kong are in the succeed-alone category, while, at the other end of the scale, countries such as Ethiopia, India, Bangladesh, Uganda, and Haiti are unlikely to raise more than a small fraction of the economic aid that they really need.

More and more, then, the emphasis is being thrown on self-sufficiency among nations, especially in food production. And that is bad news for wildlife. Increased population means increased

The vast Amazon Basin is one of the world's last remaining wildernesses. Now the Trans-Amazonica highway will cut a 3300-mile concrete swath through the heart of it, destroying part of a forest that has endured perhaps tens of millions of years. But the greatest casualties are the Indians, whose forest culture has no hope of surviving the many influences of industrial civilizations.

pressure on land resources. Any wilderness capable of cultivation, however marginal, becomes a candidate for plowing. Wetlands are drained, forests cleared, grasslands burned and reseeded. The world's great wildernesses are sure to shrink to preserved vestiges in national parks, as has happened over the last few centuries in Europe and, to a lesser extent, in America. In mid-1975 the UN Food and Agriculture Organization warned that severe overgrazing had become chronic in no fewer than 37 African and Middle Eastern countries—a sure sign of the intense human pressure on the dwindling resource of land. This overgrazing, claims the organization, is turning 8 million square miles of once-fertile land into an arid region bigger in extent than the combined Sahara, Australian, Arabian, Gobi, and Kalahari deserts.

Where pressure is so acute, even national parks can scarcely be expected to survive. In Ethiopia, for instance, it is hard to imagine that farming and grazing will not continue to encroach upon the Simien Mountains, home of the unique walia ibex and the gelada baboon. If a world recession led to a decline in tourism, the pressures to open East African parks to ever-greater human settlement would become hard to resist. The national parks of the Indian subcontinent, of Southeast Asia and the Philippines, and of South America—to say nothing of the undesignated wilderness that still survives in those areas—are all under threat. What is now being done to open up the Amazon rain forest is a case in point. It neatly displays the current and future problems of man as well as of wildlife.

When complete, the Trans-Amazonica highway will run from the Atlantic coast near the easternmost point of South America clear across Brazil to the Peruvian border. It will stretch approximately 3300 miles, running roughly 300 miles south of the Amazon itself, through the heart of the largest tropical rain forest in the world. Its effect on the forest is likely to be less than was at one time feared, for the land is proving unsuitable for any kind of farming, even ranching, and the forest is therefore unlikely to be encroached upon at any distance from the highway. The only real threat comes from the discovery of minerals, which could open up ugly pockets of development. At the moment, however, it is the forest Indians who bear the brunt of the attack.

The notorious extermination campaign deliberately waged by Brazil's misnamed "Service for the Protection of the Indian" has been halted (though casual killing of tribes that are "in the way" of ranchers, farmers, and mining companies still goes on). But everyone who has studied the problem realizes that the end of the forest Indian's way of life is only a matter of time. Sympathetic treatment could prolong the time and reduce the shock experienced by any single generation, but nothing can preserve the old ways indefinitely, short of a high-voltage fence around a few hundred square miles of forest and a total ban on contact between the outside world and the preserved "savages." The chief threat to the Indian comes not from the negative effects of such contact (drink, epidemics, cultural shock, etc.) but from the positive ones, especially from modern medicine. Recent studies have uncovered a small population explosion in villages where measles, diphtheria, internal parasites, and other health hazards have been overcome. In one village, for example, half the population was under the age of 10 at the time of the study. Population growth like this is enough to threaten the survival of forest bands living at subsistence level. The attractions of civilization, especially for the young, will complete the work. There you have a microcosm of the problem of mankind today.

The Indians have been learning about the forest ever since they were driven into it by the first Portuguese explorers and settlers of Brazil. Let us hope we can pick up some of their forest lore before it vanishes forever. The loss will be ours even as we gain a new highway. It is just one example, however, of a kind of loss that extends over the whole of the living world. We human beings live by improving on nature, but the "improvement" gets defined in our own narrow terms; and a narrowly defined improvement for man usually represents a distinct loss to the world in general. Consider the fate of what was once a huge diversity of crop plants.

Diversity is a nuisance to the farmer. It means that his plants grow to different heights and ripen at different times. It means that progeny bear little predictable similarity to parents. It brings randomness and disorder where the farmer needs regularity and predictability. Agricultural progress, leading to high yields and great efficiency, has largely been based upon achieving the genetic uniformity that does not exist in nature. Progress comes as a result of careful selection of parental stock for desirable features such as high yield, disease resistance, flavor, keeping quality, efficient conversion of the nutrient input, and so on. Then comes further selection among offspring and *their* offspring until a predictable strain emerges. During this process, the breeder inevitably eliminates a large number of variant genes from his breeding stock, until the variability of the final crop, herd, or flock is possibly only one hundredth or one thousandth part of the variability of the organisms he started with.

The resulting situation is basically unstable. It needs constant vigilance and perpetual readjustment by its human creators to keep it going. By contrast, the very diversity of the "inefficient" natural ecosystem gives it strength. If a given organism is suddenly faced with a new pest, a new predator, or an old one with a new trick, its numbers may decline drastically for a while. The more genetically diverse it is, however, the more likely it is to have among its population a few individuals that are genetically endowed to win out over the new threat. That is a rule of the survival game.

It is a rule that the agricultural breeder has to break in order to get his high-yielding, genetically uniform plants and animals. Take, for instance, the development of the "miracle" wheat and rice that were once thought to hold out such hope for a starving world. These high-yielding cereals were developed a few years ago, from natural strains that showed unusual ability to take up nitrates and build large heads of grain. This made it necessary for the stalk to be short and stiff, so as to be able to support the extra weight. Out of literally thousands of different varieties of wheat and rice and their wild progenitors around the world, only a handful have such stalks. Miracle rice (known as IR-8) was bred from a dwarf rice found only in Taiwan. Miracle wheat came from a short-stalked variety called Gaines wheat, which grew in the northwest of America.

Between them these two varieties represent a small fraction—perhaps (and what a frightening thought this is!) the smallest *viable* fraction—of the total genetic diversity of the world's wheat and rice species. In one sense they are the fittest cereal for our needs: they fill empty bellies, and turn subsistence farms into profitable enterprises. By the short-sighted logic of modern farming, they "deserve" to replace other varieties.

Well, they are doing that, all right. Africa, once a center for new rice varieties, has now lost many hundreds of irreplaceable native strains.

Twenty years ago, most of the rice in the Upper Volta plain was African in origin; today, more than 90 per cent consists of the Asian kind. In the past 15 years, thousands of varieties of native wheat have died out in Iran alone, as we happen to know from careful studies done there. We can only guess at how much we must have lost elsewhere without even knowing we had it. The Middle East and Asia Minor are the original home of all cultivated wheat varieties. In Turkey, the wild species from which cultivated wheat was bred now survive only in graveyards and ruins.

Other crops, too, have sacrificed variety to high-yield uniformity. Again in Turkey, for example, in the Cilician plain, not a single local variety of flax survives, even though this is the homeland of many cultivated varieties. The same thing has happened in Argentina, where only the cultivated strains of flax now survive. South and Central America were the original home of Indian corn, or maize. Fortunately, a few people awoke in time to the threat posed by improved strains and hybrids and in the 1960s began collecting wild strains before they could all vanish. Even so, several have been lost for ever, and many more are hard to find.

You may think that I am making too much of the danger in such losses. After all, cultivated, improved, and hybrid varieties have been replacing original strains for thousands of years without any attendant disaster. Perhaps in those millennia we have lost tens of thousands of wild and primitive domestic varieties; perhaps it is only our newfound ability to *record* the loss that makes the process seem so alarming.

Unfortunately, such comforting arguments fail to take account of the rate of change, which is all-important. Suppose, for instance, that we begin with a large tract of disease-free land and sow it with a new variety of rice. Suppose, also, to take the most favorable case possible, that this variety is totally resistant to all existing pests and diseases of rice. Then suppose that, alongside this huge uniform tract, we plant a farmer's nightmare: an indiscriminate mixture of all the strains native to the area and all the cultural varieties developed from them throughout history. How will these two tracts fare?

It is an Achilles-and-the-tortoise race, with the miracle rice spurting ahead at first like Achilles, and the hodgepodge tract playing the tortoise, giving a constant moderate yield although plagued by pests and disease. But there is such a

The Man with the Miracle Rice

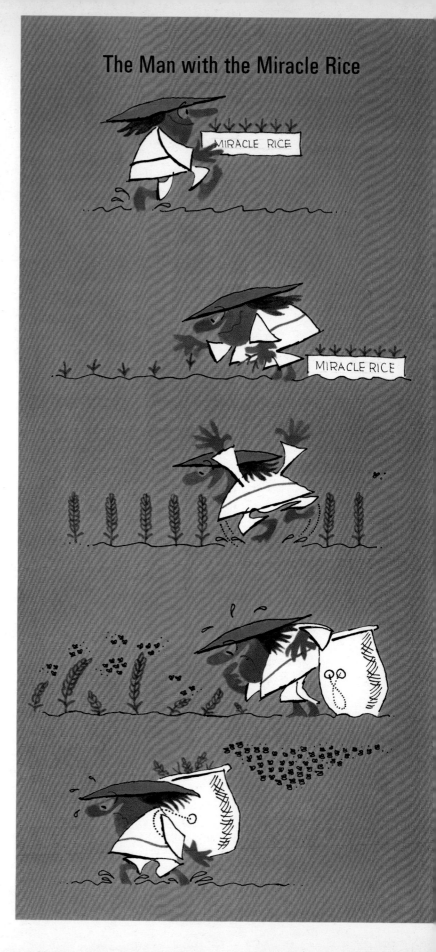

Miracle rice has a high yield and is genetically uniform: that is, each individual plant will behave identically in identical circumstances. Let us assume it starts out with a high resistance to all the known pests and diseases.

The farmer plants it in high hopes. Given proper care it can more than double his yield per acre, turn him from a subsistence farmer to one with a surplus to sell, which will give him cash to spend on further improvement.

Early results live up to all his expectations—a miracle yield. But a new pest, against which the new rice is defenseless, is no less delighted with the crop. For the pest the miracle rice is a bonanza of unparalleled magnitude.

Because every plant is genetically identical to every other plant in the miracle tract, not one of them has the power to resist the new pest. All of them succumb to its onslaught, and the miracle takes on a tarnished look.

It's think-again time, or time for miracle 2 to replenish a harvest that has been disastrous from every point of view except that of the pest. To find the genes to resist the pest calls for the resources of a rice stock that is genetically diverse.

The Farmer with the Hotchpotch

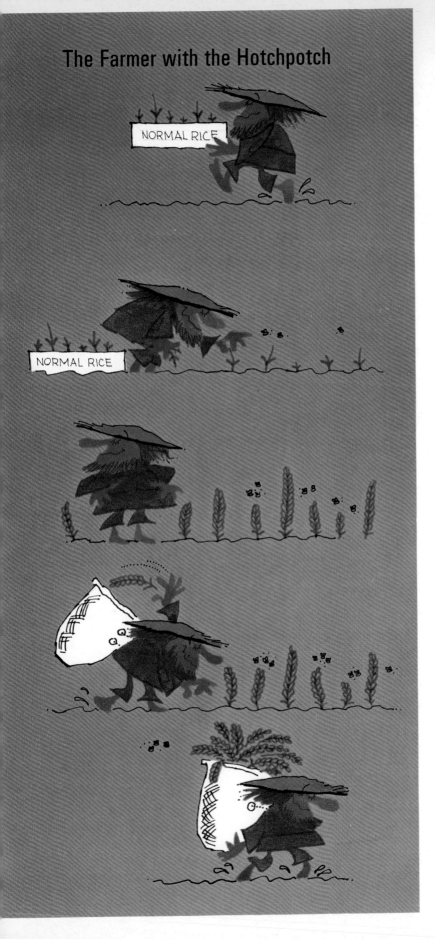

NORMAL RICE

It looks bad for this fellow. His rice is such a mixture of strains and types. They do not ripen all at one time. They do not yield uniformly. And their resistance to the many pests and diseases that afflict rice is very uneven. Still . . .

NORMAL RICE

Here is the moment the pests have been waiting for. There is something for everyone here. If one plant resists them they can still find another that is defenseless. Their depredations make growth and yield more ragged still.

Oh yes! The fellow with the miracle rice has the edge here all right. Compare his beautiful, uniform, high-yield paddy with the patchy, unkempt-looking plot of the hotchpotch farmer. Time to get rid of all these different strains of rice.

Or is it? The farmer with the miracle rice does not look so happy now. His uniform genetic strain has gone down uniformly before the invading pest. Despite his very patchy yield the farmer with the hotchpotch has more rice.

There is a lot to be said for diversity. The breeder of miracle rice 2 (and 3, 4, 5 . . .) is going to have to seek among the old hotchpotch varieties for strains that are naturally resistant to pests and diseases. The hotchpotch is our best insurance against disaster.

American Indians once lived in ecological balance with the herds of buffalo they hunted (above). White men, however, hunted the species almost to extinction. Recently the buffalo's ability to gain weight on grass alone, needing no expensive corn feed, has been used by one California breeder to make a cross with beef cattle (right). But if the buffalo were extinct, where should we now find this valuable ability?

thing, remember, as an Achilles heel—a fatal weakness. For our miracle rice, this fatal flaw lies buried in the comforting half-truth that it is "totally resistant to all existing pests and diseases." We tend to forget that a disease or a pest is just as genetically diverse as anything else in nature. A plant that is resistant to a pest is really resistant only to the particular strain of the pest that happens momentarily to be dominant. Somewhere among the genetically diverse pests and diseases that afflict rice there will be—by accident—a lucky few that are not repelled or overcome by our miracle strain's resistance factors. They are its fatal weakness.

For a year or two the miracle rice fulfills everybody's dreams. Then in one or two patches we notice that a pest or a disease seems to be taking hold. It does not really worry us at first because the yield of the tract is still great, but after 10 years or so the situation has become disastrous.

The pest has at its disposal a vast tract in which every single plant is defenseless. Meanwhile, in the other tract the steady, unspectacular yield continues unabated. It goes on and on, whereas the miracle rice has become a museum piece within 15 years, able to survive only where it can be carefully hand-tended.

Long before that, if we have any sense, we shall be hybridizing a resistant wild strain and the susceptible miracle strain. Moreover, we shall have started a regular hybridization program to anticipate future troubles with pests, and so to meet them halfway.

That brings us back to the importance of the rate of change. It takes 15 years for miracle rice I to become a museum piece; it also takes 15 years to produce miracle rice II, to test it, to breed the seed, and to get it out into all the paddies where miracle rice I used to flourish. If, on a world scale, we permit an interval between the production of rice I and rice II, millions of people will starve to death. The time to start producing rice II is when the air is still ringing with self-congratulation at the first bumper harvest of rice I. But that is if the pests and diseases follow the normal, rather slothful 15-year diffusion pattern. If we want to guard against really virulent pests and diseases, the time to start work on miracle rice II is about a decade before we even know what number I is going to be. Which is another way of saying that we must sacrifice some of the miracle to be certain of keeping the rice.

We had our timing about right in the earlier part of this century, when plant and animal breeding was more nearly an art than a science. Hampered by their ignorance, breeders achieved their miracles more slowly, but not less surely. The improvement in the yields of all kinds of food plant and animal that took place between, say, the 1880s and the 1910s was certainly a miracle; but it took place at a rate that let breeders adjust to the corresponding adaptations among pest-disease populations. Thus, the growers stood more chance of consolidating their gains as they

made them. Today, hampered by a little learning, we improve our Gaines and our IR-8, and then fight desperately to conserve them.

Sensible hybridization based on diversity remains an area of hope for future food supplies. There has been a recent ray of light in the animal world, for instance, where rocketing grain costs have hit beef farming badly. Most beef is reared on grass but fattened just before slaughter on expensive grain, usually corn or barley. To cut costs, a California rancher, Bud Basolo, has experimented with crosses of buffalo and beef cattle. The resulting hybrids ("beefaloes," of course) are fertile. Not only that, they also put on weight faster than beef, *and* they do it on grass alone. Basolo reckons that beefalo meat can undersell grain-fed beef by 20 to 40 per cent, depending on the season. Around the turn of the century buffaloes were so close to extinction in America that it needed an experienced tracker to find even one or two. Suppose that no serious conservation program had been undertaken and that we had pushed the buffalo just that bit further into extinction. Where should we now find the genes to build beef on grass alone?

The most forward-looking answer to that question may well be: "Manufacture them!" In other words, one of the most realizable hopes for the future lies in genetic engineering—the deliberate creation of mutants, with characteristics that are tailor-made for special functions. This has long been a dream of science-fiction writers, but events over the past quarter of a century have brought the cherished dream quite close to realization.

As long ago as 1865, an obscure Moravian monk, Gregor Mendel, discovered that hereditary characteristics were passed on through the generations in undilutable packets that he called "genes." He showed, for instance, that greenness and yellowness in peas is determined by genes, and that if a pea inherits a greenness gene from one parent and a yellowness gene from the other, the yellowness gene takes a back seat, letting the greenness gene express itself fully in the offspring, which thus has the appearance of a pure green pea. The yellowness gene still perseveres, however, in the offspring's genetic makeup. And when each of the pea's pollen-producing cells starts to divide, in order to make two pollen grains or two egg cells, the genes separate as well. The greenness gene goes to one, the yellowness gene to another. So an apparently pure green parent can make pollen and eggs that contain the instructions for making a truly all-yellow offspring. All that is needed for this is for a yellowness-bearing pollen to fertilize a yellowness-bearing egg.

In short, Mendel showed that there is an elusive "something" in hereditary material that can survive undiluted through the generations. It took almost a century of hard work on the part of thousands of biologists to determine exactly what that elusive something is. Ironically, it proved to be DNA (deoxyribonucleic acid), a substance first discovered in living cells in 1869, a mere three years after Mendel first published his work.

The honor—and a Nobel Prize—for discovering the role of DNA went to two men working at Cambridge University in 1953: an Englishman, Francis Crick, and an American, James Watson. From X-ray pictures taken by another English-man, Maurice Wilkins, who shared the Nobel Prize with them, they deduced that DNA, which is found in the nucleus of all living cells, has a molecule shaped like a double helix—that is, like two identical interlocking spiral staircases. They figured out, too, that the spirals are made up of only four basic building blocks, which they labelled A, C, G, and T, and that A is always paired with T, and C with G. This accounts for the twin spiral, for wherever A occurs on one spiral, there is a corresponding T on the other;

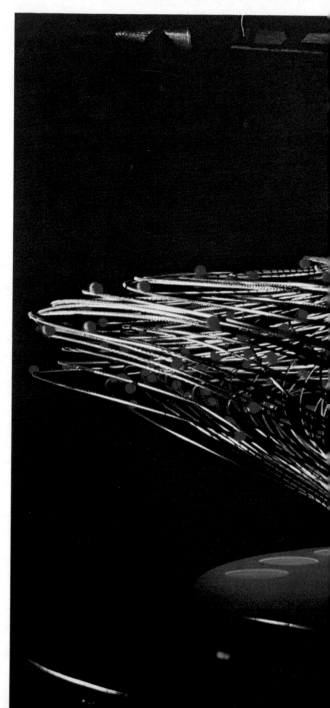

A large-scale model of a section of a chromosome. It is in the threadlike chromosomes found in cell nuclei that the genetic code that controls the life processes in plants and animals is found.

and similarly with C and G. This pattern of correspondence keeps the two spirals together—like a zipper coiled around a long tube.

The mechanism by which DNA transmits hereditary characteristics from one generation to the next is too complicated to be effectively summarized in these pages. It is enough to say that the infinitely variable arrangements of A, C, G, and T on their partnered spirals form a code that specifies the building of protein according to a certain pattern; and when a cell divides in the normal course of growth, the double-spiral DNA molecule unwinds and divides into its two halves, each of which recreates its missing partner. However, when the DNA is, so to speak, unzipped for the division, it becomes vulnerable to slight changes, and even a very small change in the DNA can result in drastic changes in the protein it specifies. New proteins, in turn, can markedly alter the nature of the living thing that makes them. This, after all, is the mechanism behind the long story of evolution—a mechanism that has

operated entirely through random changes and their interactions with the environment. Random though it is, a mechanism that took the rather elementary DNA which must have existed in primeval microorganisms and from it evolved the incredible variety of life as we know it today is clearly a very powerful and versatile force.

Since we now understand the mechanics of the genetic process, why should we leave the production of further variations to chance? Why should we not use our biochemical knowledge to create deliberate changes in the DNA of living things, and open up a wealth of technical possibilities?

In fact, such genetic engineering is already under way. For some years now, the people who study DNA and its mechanisms in research laboratories have routinely introduced deliberate changes in the simple DNA of certain harmless bacteria. In many countries commercial as well as academic researchers have got into the act. Understandably, they are still working entirely with the DNA of the simplest microbes, which reproduce quickly and so show the results of experiments within hours; also, to be sure, microbe DNA is much easier to understand and to change than would be the DNA of more complex organisms. Even with the microorganisms, however, the potential rewards are dazzling to contemplate. For instance, a modified bacterium that could leach out copper, gold, or other valuable metals from low-grade ores would transform our gloomy resources forecasts. Thousands of millions of years of random mutation have not produced such a creature (not one that has survived, anyway), but perhaps we can do better with deliberate mutations. Another line of current research is aimed at taking the gene that permits multicellular animals to manufacture the hormone insulin (lack of which causes diabetes in humans) and incorporating it intact into a unicellular microbe. Such a mutation would, of course, provide a route into the cheap manufacture of commercial insulin. Among some other fairly obvious candidates for genetic engineering are microbes that could digest various kinds of waste matter and synthesize such useful products from the wastes as ethanol (a gasoline substitute) and the simple raw materials that are the starting

The cow of the future? Essentially an udder attached to vestiges of the other organs needed by traditional cows—a living machine. Actually, this one is a product of photographic rather than breeding skill; but an ambitious genetic engineer of the future might set himself just such a goal to aim at.

How to get a dozen average-quality heifers to bear championship calves. First, champion heifers are given a fertility drug to induce the shedding of many eggs for fertilization. Then they are artificially inseminated with champion-sire sperm and, after a brief incubation in the mother, the embryos are flushed out (left). Each is surgically implanted (right) into an ordinary heifer on heat. There the champion embryos grow to full term while their biological mother is immediately available for further service. Such techniques herald the dawn of genetic engineering — the deliberate alteration of natural biological events.

point of many chemical and industrial processes.

Probably some time after the beginning of the 21st century, our almost certain eventual success in these early experiments could pay off in more daring raids upon the DNA of complex, multi-celled organisms. Here, too, the possible benefits could be vast. Take agriculture, for instance. Most cattle breeders agree that we are nearing the limits of the genetic capability of the milk cow; if we want to improve her yield, productivity, butterfat percentage, or anything else, we are unlikely to find an available gene in the existing "gene pool" (the total DNA of all living milk breeds). We must, therefore, either wait for a one-in-a-million lucky mutation to occur naturally in the DNA or learn to modify the DNA along the desired lines. The second alternative is becoming an increasingly likely possibility. Two centuries from now, one type of cow could well be a living milk-making machine in which all other functions (and the organs that go with them) will have been suppressed or reduced to vestiges of their present size and capability.

Similarly, an apple tree is a very inefficient way to make apples. Most of the energy that the tree soaks up from the sun goes into producing valueless timber. Perhaps the orchard of the future will be filled with "trees" no larger than strawberry plants, each bearing a handful of full-size apples.

But the dangers that would accompany such advances are as immense as the dreams. For one thing, there are already plenty of harmful bacteria and viruses in nature against which we have so far found few or no defenses. The possibility that we might add to their number, albeit unintentionally, with man-made creatures unknown in nature, and therefore of unknown virulence, is terrifying. In 1974, faced with this possibility, a panel of 11 leading US researchers took a step without precedent in the history of science: they declared publicly that certain key experiments aimed at modifying the bacterium *Escherichia coli* were being brought to a halt, and they urged their colleagues everywhere to follow this lead. Their reason was that the work that had already been done on *E. coli* was so close to fruition that the scientists immediately involved in it feared lest they might inadvertently create a novel form of the bacterium capable of resisting all known drugs and other antibacterial agents. Before

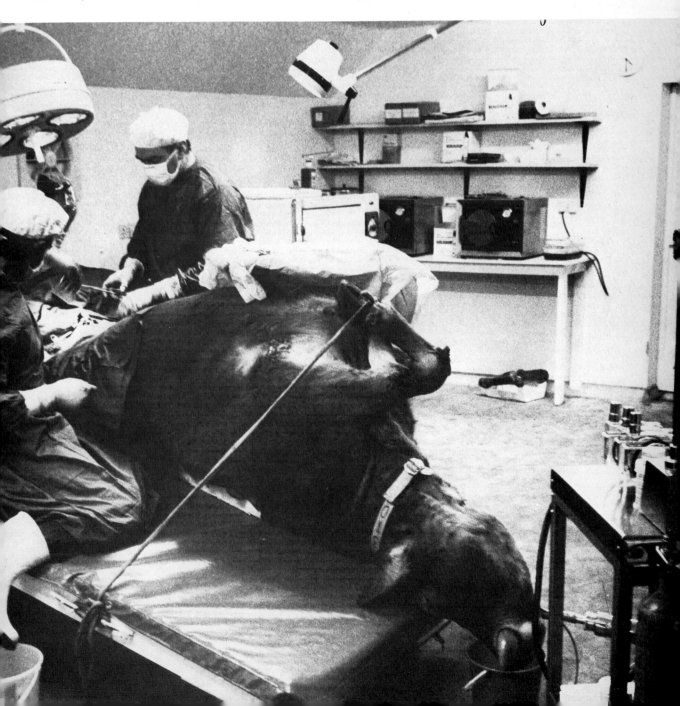

Above: the danger that genetic engineering would find military application (here symbolized by a researcher in chemical and biological warfare in full regalia) has led many of today's biologists to approach their work with great caution.

resuming the work, the scientists announced, a complete evaluation of the hazards, with no unanswered questions, would be an absolute necessity. Among the 11 men on the panel were some of the most careful and conscientious researchers in the business, including James Watson, one of the discoverers of the genetic role of DNA. Indeed, that was their point: if *they* felt that a project in which they were interested should be stopped, how much greater should be the compulsion on less experienced workers.

Escherichia coli is the most common bacterium in the human gut. Textbooks label it "harmless"—which it is, as long as it stays in the gut, although it can turn lethal when it runs rampant in, say, an accidental wound and invades other organs. Because it is harmless in normal circumstances, it seemed to be an ideal subject for experiments in DNA manipulation. One impor-

tant center for all such research has been Stanford University in California, and it was there that a team led by Dr. Paul Berg was developing bacteria-invading viruses as "ferry boats" to carry new bits of DNA into bacteria and to attach this new material onto the bacteria's own DNA. Berg and his colleague Dr. David Jackson had perfected an elaborate chemical technique for achieving this switch and had carried out each stage separately, but they had never put the whole sequence together and actually modified an *E. coli* bacterium.

The experiment they proposed was, in all conscience, modest enough. They simply wanted to try to modify the bacterium's way of digesting a particular kind of sugar. They had the new DNA code worked out. They had the DNA itself, perfectly synthesized and attached to the right kind of virus, ready to invade the bacterium. But there they paused. Ever since the start of this research they had been uneasy about the possibility that something might go wrong—about the risk of unwittingly creating new bacterial and virus diseases. For a long time, too, Berg had been critical of other scientists who, in his view, were not sufficiently careful about setting up safeguards against such dangers. So it is not surprising that Berg chaired the panel of scientists that decided to declare a moratorium on the genetic engineering of *E. coli*.

A conference of leading genetics researchers held in California in 1975 did not entirely back the call for a complete—though temporary—halt to genetic-engineering projects, but it did endorse the view that certain specific areas of research were too dangerous to be fully explored at present. These sentiments have since been expressed by other groups. No such affirmation, however, has the force of law. And that is where the matter now uneasily rests, with nothing but moral exhortation between the world and projects "too dangerous" to be pursued "at present."

Whether you regard the trends revealed in this chapter as gloomy, neutral, or hopeful, depends more on your own temperament than on my summary of the situation. The trends are so many, so varied, and so uncertain in their onset, course, and outcome that it would be absurd to attempt an evaluation of their overall impact. Yet, as we shall see, our hunger for an overall rating, an overall trend, an overall prediction has exercised a powerful—and powerfully misleading—influence on today's futurologists.

Below: in H. G. Wells's The War of the Worlds *the Martian invaders are quelled by bacteria against which they have no immunity. Genetic engineering might create bacteria against which we, like the "Martians," would be defenseless.*

Some Perspective

From our brief survey of the young science of futurology, its findings, and their implications for us and the living world in general, we have become aware, I hope, that basically there are two different ways of looking at the future. It should be obvious, too, that very few people keep the distinction between those two approaches clearly in mind when discussion turns to the future of this, the future of that, or the future of things in general.

One kind of future is hard to discuss, because it will really happen—and we cannot foresee all

A model for predicting catastrophe. So many things in nature and human affairs happen suddenly—slumps and booms occur, wars begin, people lose their temper, TV pictures break up. In each case a slight change can make a catastrophic difference. France's great mathematician René Thom has devised theorems to cope with this seeming capriciousness. The elegant curve below is a 3-D expression of formula explained on page 122.

the twists and turns of reality. The second kind—the sort of projection that tends to obsess us—is the if-things-go-on-as-they-are kind. First we look at the growth of population, or the rate at which we are consuming our resources, building nuclear generators, paving over farmland, polluting the oceans, or doing any one of a dozen other ruinous things. Then we make some necessary mathematical equations, draw an exponential curve, and "prove" that the world will go out of control before the year 2010.

Any such prediction assumes that the future is a direct extension of the present, that no surprise developments will intervene to change the course of events, and that no activity will tail off as it reaches some point of saturation or satisfaction. For example, the straight projection assumes that *everybody* will work as hard to buy a third car as he did to get his second, and that he will then go on slaving without letup for his fourth and fifth cars.

With that brand of ignore-the-real-world mathematics, and on the strength of hindsight that uses historical data from the previous

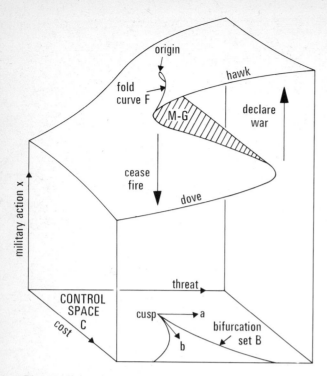

Diagram of the sculptural form on the previous page. It describes sudden, or "catastrophic" changes in certain systems. A country with a peaceful attitude (dove) is increasingly threatened until it declares war—the extreme of the hawk position. As the threat is reduced, there comes a point where peace is suddenly declared. The "overhanging cliff" fold allows for sudden changes of this kind. Such models have made little impact as yet on the mathematical side of futurology but are obviously destined for an important role.

centuries, we can easily "prove" Manhattan sank under the accumulation of horse dung around the turn of the century, or that England's canal builders removed the last square foot of earth around 1890 and the country became one continuous sheet of canal water. (Meanwhile, we can also "prove" through an alternative set of past statistics that the entire land surface of every industrialized nation became covered by a vast network of railway tracks.)

As we know, all such projections—if anybody had made them—would have been wildly awry. Even carefully corrected projections, which would have taken into account the fact that there comes a time when new canal and railroad building starts to fall off, would have been extremely wrong. Sensible though such projections might have tried to be, they could not have foreseen the enormous effects of such surprises as the automobile and the passenger-carrying airplane. It is always the surprises that make the actual future different from even the most thoughtfully reasoned-out future. The prediction

of a "surprise-free" future is almost certain to be gloomy. It assumes that the growth of anything will be exponential; and, as we have seen, limitless exponential growth is bound to reach a point where raw material, land, clean water, or any other essential substance is all used up.

The best-known modern projection of this kind appears in a book called *Limits to Growth*, published early in the 1970s, by Professor Dennis L. Meadows and a team of co-workers at the Massachusetts Institute of Technology. The authors' basic proposition is simple: on a planet where space and resources are limited, growth cannot possibly be infinite. This self-evident truth lends a great deal of weight and appeal to the rest of their thesis, which is that the limits to growth here on earth are already very near. In fact, says the book, we have so little leeway left that, like an express train headed for the buffers at the end of the line, we ought to apply the brakes right now to prevent disaster.

The main grounds for this conclusion lie in the exponential growth of a number of critical factors in world civilization—especially, of course, the factors of population and industrial output. We saw in an earlier chapter how small is the annual increment (the new amount added each year) in the first phases of exponential growth—but how in the later phases it becomes astronomical. This is, indeed, what is now happening to population and industrial output, and *Limits to Growth* maintains that unless we take some immediate and effective action, each annual increment will inevitably generate an increasing instability in the world system.

The constraints within which we are trapped are the finite amount of cultivable land, the finite stock of nonrenewable minerals, and the limited capacity of the biosphere to cope with pollution. Although modern technology can ease some of the constraints, it cannot ease all of them, and it cannot help us to hold out indefinitely. Sooner or later (and the MIT men maintain it will be sooner), the express-train growth of population and industrial output will crash into the inevitable buffers, with catastrophic consequences. Unless something is done at once, say the authors of *Limits to Growth*, there will soon be a decline in total world population and in industrial capacity.

Our only hope of salvation, as they see it, is to recognize the constraints now instead of letting the future impose them upon us violently. Specifically, we must establish policies of zero growth in population and stable industrial output during this present decade. Professor Meadows and his colleagues concede that their data can be interpreted as allowing slightly more optimistic assumptions than their own. They insist, however, that in the long run it makes no difference; the most optimistic outlook merely postpones the crisis without changing its character.

The inevitability of the crisis, say the authors, comes out clearly in the predictions arrived at by feeding data into a computer programmed with a model of the world system. Their model is actually a set of mathematical equations of the kind we looked at in Chapter 2, but it can also be expressed as a diagram that shows the complex relationships among five major aspects of the world we live in: resources, population, pollution, capital investments, and agriculture. The program was run and adjusted several times, until it showed a plausible agreement with trends as we know them to have occurred between 1900 and 1970. Then the MIT team ran it forward, without changing any of its basic assumptions, as far as the year 2100. The results as plotted in a series of graphs show a drastic fall in food supply per person and in available natural resources from now on. Because of the inertia of the system, though, both population and pollution continue to rise for a while—until famine, disease, and the total collapse of industry inevitably bring them to a halt.

Of course, as the authors emphasize, this is not a prediction of the "actual" future; many things can happen to disturb a smooth, standard, computerized projection of trends (the "standard run", as it is called). We can discover new resources, raise soil fertility, breed more productive plants and animals, and curb pollution; or, as the price of a commodity goes up, we can learn to use it more economically or to recycle it profitably; or birth-control programs can begin to work or fertility patterns to change; or starvation can hit disproportionately at the young, who are the breeders of the next generation, so that population estimates alter drastically. And so on. Ghoulish though one or two of these eventualities may sound, *all* of them would be "favorable" in that they would postpone the catastrophic collapse. But they would not prevent it. Among the most interesting of the *Limits to Growth* predictions are those in which certain favorable assumptions were fed into the computer in order

to discover how they would change the curves.

These alternative computer runs show that if the assumptions on which the model is based are correct, partial solutions to the big problems do not in fact solve anything and often land us in a worse mess than we would be in if we simply left things alone. For instance, even if we controlled pollution, increased agricultural productivity, discovered virtually unlimited resources, and instituted the most thorough birth-control program possible, the crash would be staved off, the computer tells us, only until late (instead of early) in the century. Indeed, the only possible way to avoid disaster is to limit the investment of new capital to the amount lost by depreciation—thus cutting all economic growth to zero—and to do this immediately, before the end of the decade. Then, according to the MIT findings, we should achieve a steady-state world: steady population, steady production and consumption, steady agricultural output, and a steady and biologically acceptable level of pollution.

Limits to Growth has had such widespread publicity that many people may believe in the book's assertions without knowing exactly where they came from, and may assume that they are part of the common intellectual currency of our time. That, however, is by no means the case. The book has been criticized on numerous grounds by a variety of experts and expert bodies, ranging from a hard-hitting (and highly readable) report from the World Bank to a detailed and technical dismantling job—published under the title *Thinking About the Future*—efficiently performed by the Science Policy Research Unit of England's Sussex University.

Some of the critics have failed to understand the highly generalized nature of the *Limits to Growth* model. Meadows and his co-workers have never claimed to be predicting an "actual" future; their model, they freely admit, is far too generalized to yield actual dates and actual quantities. They assigned no physical quantities apart from dates to their graphs precisely because they did not want discussions to be bogged down in details. Instead, their aim was to get a picture of the general behavior of a dynamic system (the world system), given different assumptions about such important variables as population, growth, and agriculture. So any criticism of details is to some extent misplaced. Some of the most devastating criticisms of the book, however, have been directed not at its details, but at the general—the far too general—nature of the model itself.

The trouble with the model is that its apparent thoroughness and complexity make it seem so impressive. It *must* be true, or close to the truth, thinks the dazed layman. Because the *Limits to Growth* process is rather typical of such ambitious and highly respectable processes—because they all have such an air of infallibility that even the expert mind tends to surrender to their persuasions—let us take a brief look at some of the more important defects.

To begin with, if you did not know better, you might think that what the model calls "resources" covers everything from oil to manganese nodules. It does not. As far as the model is concerned, there is just one aggregate world resource, and it gets consumed by just one homogeneous aggregate population, fed by a single aggregate agricultural system, creating a single abstract pollution that stands for bacterial pollution, organic wastes, dumped heat, and radiation—all somehow brought together in a single entity. The MIT men claim that because these single aggregate entities are based on figures that go back over the years 1970–1900, they can be relied on, despite the vast amount of abstraction required in order to fit them into a manageable computer model. In fact, though, it is dangerously misleading to average quantities out in this way, even over such a long span as 70 years—and it can be especially misleading if the quantities look as if they are going to zoom off on an exponential curve.

For example, consider the sequence of an aggregate quantity (let us call it X) whose average values at 10-year intervals are: 100; 1604; 2364; 3528; 5312; 7968; 11,450; 13,600; and so on. That represents exponential growth at a rate of close to 4 per cent a year for 70 years. But suppose that X is, in fact, an aggregate of three quite different quantities, P, Q, and R, and that each of these quantities behaves as follows over those same 70 years:

P: 1000; 1409; 1985; 2799; 3945; 5559; 7834; 11,040
Q: 100; 197; 387; 762; 1500; 2951; 5808; 11,430
R: −0.5; −2; −8; −33; −133; −542; −2192; −8870

Now if we run the aggregate X ahead for a further 10 years, we get a figure of something like 20,000. But note what happens if, instead, we project each of P, Q, and R separately: P becomes 15,560 after another 10 years; Q becomes 22,490; and R becomes −35,890. In fact, the true aggregate is only 2160, very far short of 20,000.

The reason for this gigantic discrepancy is, of course, that X itself does not exist. It is merely a convenience total, which saves us the chore of working separately with P, Q, and R and the obviously very different equations that govern their growth or reduction over increasing time spans. And in the real world aggregates of the kind used in the *Limits to Growth* model are composed not just of three quantities but of dozens or even hundreds—something quite beyond anybody's ability to model or any machine's power to compute.

Thus, it is quite possible, from watching the fairly consistent behavior of an aggregate quantity such as X for 70 years, to come to the reasonable conclusion that it is going to shoot out through the roof in the foreseeable future, *and to be entirely wrong*. When the real components of X—P, Q, and R—are examined, we begin to realize that such situations, even if predictably consistent, are much more complex than they might appear at first sight to be.

If you look at the fine details of the *Limits to Growth* model, it soon becomes clear why it will predict collapse under any conditions except that of zero economic growth. Bleak pessimism is implicit in the way the model apportions the world's income and its expenditures. Specifically, in apportioning how much of our industrial output is consumed and how much (in money terms) is plowed back into new investment, Professor Meadows and his team assume that at all times and in all conditions 57 per cent is plowed back, and 43 per cent is consumed. Why? How did they arrive at these figures? The answer is that in 1968 (one year only!) the data for 54 countries (hardly a full representation, because none of them had socialist economies) averaged out at just those two percentages. Actual figures for the 54 countries showed a wide spread—and the percentages were not markedly related to national income, stage of development, or type of economy.

So the 57/43 split is based on an incomplete and nonrepresentative collection of data. But it is not this alone that forces the model to predict catastrophe in every circumstance except that of zero growth. There is also the question of how the plowed-back 57 per cent of industrial wealth is divided among such segments of the world's economy as services (banking, shops, administration, etc.), agriculture, and industry itself. The MIT study assumes that when *world* income

The *Limits to Growth* World Model

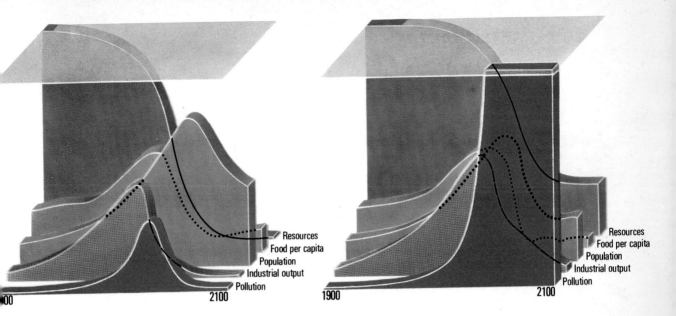

2100		1900	2100

Labels (left graph): Resources, Food per capita, Population, Industrial output, Pollution

Labels (right graph): Resources, Food per capita, Population, Industrial output, Pollution

The Standard Run

Unless there are important changes in world developments, the Limits *model predicts runaway growth followed by collapse around the year 2050. Piecemeal tinkering with the system— as the next six graphs show—does not fundamentally change the dynamics of the system or the long-term consequences.*

With Resources Doubled

For instance, if we discovered double our present resources, industrial output would surge ahead and lead to a pollution crisis around 2020. This, in turn, would cause widespread death and drastic falls in food production. Resources, despite being doubled, end up severely depleted.

is, say, $400 per person, the reinvestment will be apportioned in exactly the same ratio as it was apportioned in 1968 in the *countries* that happened to have national incomes equal to $400 per person. On that assumption, the way to find out how the whole world will be reinvesting its wealth when world income rises to $1600 per person is to consult the 1968 charts to see what countries with a $1600-a-head income were doing back then. Moreover, *Limits to Growth* assumes that reinvestment in services and agriculture gets first call on the 57 per cent, with industry taking whatever remains. There is no accepted economic theory to account for this apportionment; there it is—just one more built-in rigidity that forces the model to behave in a way totally different from that of the real world.

What does industry do with its residual portion? Again there is an arbitrary set of assumptions. Part gets used up in finding new resources to replace those that have become depleted. Part

goes to cover depreciation. The rest goes to promote new industrial growth. The proportions here change drastically, however, as the years roll on. For example, in the early years, while resources are still abundant, industry spends only one twentieth of its share on seeking new resources. Later, this fraction rises to almost four fifths, so that industry is left with only enough new capital to cover depreciation, with nothing earmarked for new investment. At that point growth inevitably comes to an end—what else can it do?—and the system collapses.

Taken together, all the rigid assumptions make catastrophe inescapable, but it is programmed, not real, catastrophe. In what conceivable real world could we have a situation in which consumption steadily sops up an unchanging 43 per cent of output, and services and agriculture take a privileged first bite at the remaining 57 per cent, while industry relentlessly squanders four fifths of its residual share on seeking further

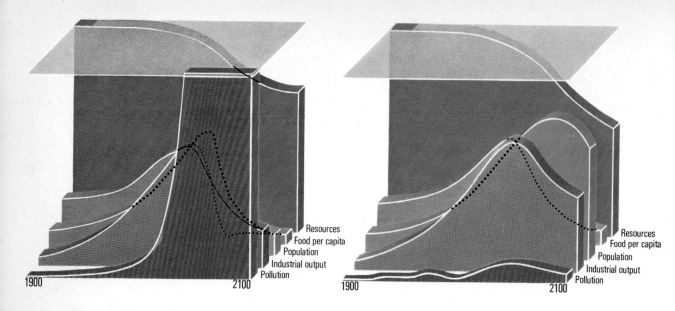

With No Resource Problem

Even when we assume virtually unlimited resources, the same pollution crisis—with the same consequences—ensues.

And With Pollution Controls

If pollution is controlled, the Limits *model predicts that we shall run out of new arable land around 2020, with a resulting population collapse around 50 years later.*

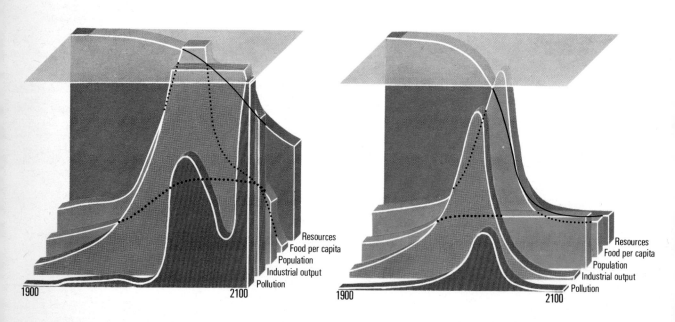

And Increased Food Productivity

If we also raise food productivity, we achieve a temporarily stable population and a world income close to the present US level per person. But rising industrial output and pollution depletes resources and causes a reduction in food output. The population will crash a century later, Limits *predicts.*

Standard Run Plus Stabilized Population Now

If we merely stabilize the 1970s world population, then, according to Limits, *industrial output leads to rapid resource depletion and increased pollution. It ends in a collapse of the industrial system and a large fall in the food output.*

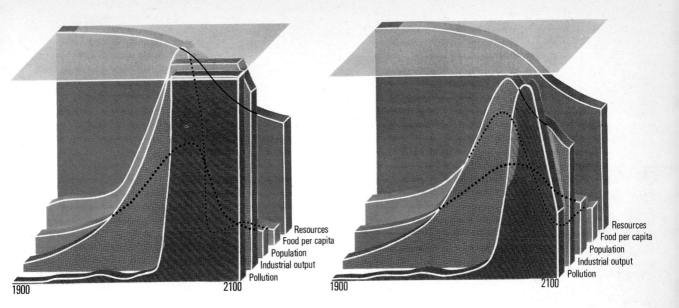

Plus Increased Food Production

If we then double the average yield of food from each acre of land, we remove so many of the usual constraints on industry and population growth that both soar ahead. Although pollution per unit of industrial production is kept low, there is so much production that pollution rises beyond the biosphere's ability to deal with it. Population still crashes around mid-century.

No Resource Problem, Good Pollution Control, Perfect Birth Control

If instead of raising food productivity, we achieve "perfect" birth control (i.e. replacement only) the preponderance of youngsters in the world of the 1970s still ensures population growth. The food crisis is postponed, Limits says, by only a decade.

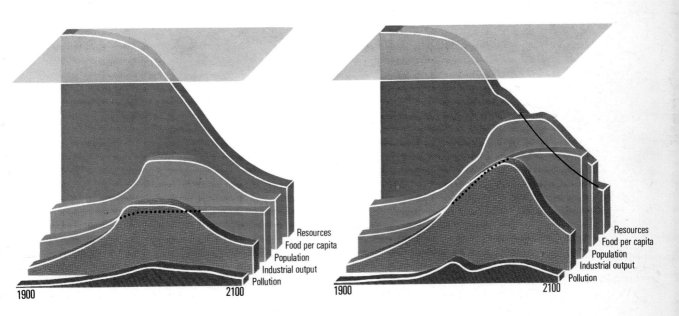

Plus Stabilized Capital Now

Only when we stabilize capital by shunning all industrial growth does the Limits model predict a stable and a viable world.

Stabilizing Policy Delayed Until Year 2000

As a sting in the tail, Limits claims that if we delay action for two decades or so, reforms that would have brought stability if they had been applied in the 1970s will then be of no avail. Food and resource shortages cause a crisis before the next century ends.

resources, leaving only enough to cover depreciation on existing capital? Yet those are the circumstances that *Limits to Growth* assumes— the very circumstances in which "collapse" is bound to occur. If real investment managers behaved like that, we could quite easily have our collapse next Saturday.

The real economy is far more flexible and resilient than the model economy of *Limits to Growth*. Like all flexible systems, it needs something to help it maintain its essential shape in spite of small deformations and stresses. That something, quite simply, is the price mechanism: when a given commodity gets scarce, its price rises and the people who need it begin to look for substitutes. Scrap materials containing the now-scarce commodity also go up in price, and so the recycling market increases. Both the high price and the increase in recycling help to conserve the raw-material resources of the commodity. Most of the economists who have analyzed the *Limits to Growth* model agree that the Meadows team has given too little weight to this most important mechanism, which does a very great deal to postpone, if not actually to prevent, economic collapse.

Any one, therefore, of three basic flaws (the use of probably misleading aggregate quantities, the rigidity that compels the model to forecast collapse in any circumstances other than zero growth, and the failure to take the price mechanism into account) is enough on its own to cast doubt on the model's forecasts. Together they help to explain why predictions such as those put forward in *Limits to Growth* are always likely to be at odds with what our own common sense tells us. Common sense tells us that nothing we do is really exponential in character. Futurologists pay tribute to this knowledge when they invent a special kind of man whose behavior *is* exponential; they call him Faustian man, in honor of Faust, the legendary German magician who, having exhausted a rich legacy, sold his soul to the devil in exchange for 24 years of unlimited power. The parallel with big-technology man, in a century that has less than 24 years of its span still to run, is strong.

Faustian man has exhausted a rich legacy of the earth's resources and now has a limitless greed for power and possessions. Give him one of anything—cars, houses, telephones, cuckoo clocks—and he craves two. Give him two, and he demands four. Faustian man is insatiable. He and, of course, his equally greedy wife are the true authors and sole subjects of every pessimistic, surprise-free future.

We know, however, that real man is not eternally Faustian. Real man works hard to get one car, one home, or one color-TV set. He may work just as hard to get a second. But as for a third and fourth—well, there comes a point when only a madman would go on behaving in the Faustian manner. If we project the behavior of *real* man back into the demand curves or activity curves that are currently causing us so much alarm, we find a very different kind of future emerging. Again, the past is a revealing guide.

Take the curve that represents sanctioned railway mileage in England from 1840 to 1846, and compare it with the curve that spans the entire decade of 1840–50. The swift growth of the early part (1840–6) seems exponential, but growth slows considerably thereafter. (One reason for the slowing down was that money for capital investment became scarce.) This type of curve, which flattens out after a soaring start, is a modification of the exponential curve; mathematicians call it a *logistic* curve, and a Canadian futurologist named John Kettle has suggested that it may be a more accurate guide to the future of economic growth than the traditionally used exponential formula. His idea is important to every living thing because economic growth is obviously an all-important factor in pollution and human impact on the environment.

What gave Kettle the impetus to start thinking about logistic curves was a set of figures published in the mid-1960s by a Canadian government planner, who had compared economic growth rates with wealth in 140 countries. To make his comparisons, the planner took the current gross national product, or GNP, of each country and divided it by the population, so as to give a measure of GNP per person. Then he compared that figure with the rate at which the annual GNP itself was growing. The result was surprising. In poor countries, with a low GNP per person, the rate of growth, as you might expect, was low. But it was also low in very rich countries, with a top GNP per person. The rate

One for the remote future, although there are already plenty of organizations and people who take the exploitation of space very seriously. Certainly high vacuum, low gravity, limitless solar power, and innumerable metal-rich asteroids suggest intriguing possibilities and a host of new problems for man.

of economic growth was greatest in the countries in between the two extremes.

An analysis of the data convinced Kettle that they conformed to a typical logistic curve—a curve, in other words, that slowly flattens to an almost straight line after a relatively brief period of arching upward. A desperately poor country, for instance, may take fully three centuries to rise from a GNP of a meager $50 a head to one of an adequate $11,000 a head. Its progress will be very slow at first; it will take a century and a half to reach $1000 a head. Then the rise will accelerate and the curve will zoom upward, with the GNP per person reaching $8000 in just over half a century. Thereafter, however, there comes a tailing off, until, at about $11,000 (in 1970 values), growth will virtually cease. This, according to Kettle's theory, is a growth pattern that may be more applicable to future global economy than many of our present-day forecasters are willing to admit.

Unfortunately, his suggestion raises as many questions as it answers. For instance, if a number of countries reach the end of the curve, does that make it easier or more difficult for those still on the way up? Does aid from rich countries promote or retard the progress of the recipient poor ones along the curve? Do free-enterprise economies move faster than centrally planned ones, or are deeper rooted forces at work, on which the organization of the economy has only a very slight effect?

Only a rather naïve economist could accept the logistic curve as an utterly dependable model for detailed prediction. Apart from everything else, the statistics on which the Canadian report was based are, like most statistics, probably inaccurate. Even so, despite all the reservations, the curve does seem to reflect past experience as well as a commonsense view of the future. To bring it down to the personal level, for example, an individual who is wretchedly poor cannot save for investment, and so it takes a long time for him to get the growth process started, if he can do it at all. Once begun, however, the process tends to accelerate as new wealth breeds further investment and greater wealth. Finally, as he becomes accustomed to affluence, Faustian man begins to lose his drive, and a new kind of man (post-industrial man?) takes his place.

The implications of his "normal" pattern are immense—and are vastly more encouraging for us and for the biosphere than runaway-exponen-

tial predictions would suggest. For instance, it means that the gap between rich and poor is likely to narrow and, ultimately, to vanish—not (as many people fear) to broaden. It means, too, that an economy based on recycling of material and on low-impact technology may actually fit the pattern of the future better than does our present-day Faustian technology. An acceptance of the logistic curve as a reasonable alternative to the exponential curve can have an important effect on our morale, too. At the moment, there is a good deal of sentiment against any effort to promote economic growth, because (on the exponential model) such promotion merely pushes more countries onto the runaway slopes to extinction. On the logistic model, however, economic growth holds out the promise of raising the whole world to a comfortable living standard, at which point no-growth stability would then become both desirable and possible.

To be sure, the smaller the world's human population at that time, the bigger our safety margin where the earth's remaining resources are concerned. For that reason, nothing in the logistic projection can be read as removing the need for action on population reduction, pollution control, recycling, and an ecologically acceptable technology. In fact, quite the reverse; the logistic projection offers more hope for the success of such efforts than does the pessimistic runaway-growth view of the future.

There is one factor, though, that no forecast based on economic projections can take fully into account. This factor—which, in the long run, may be the most important of all—is our own nature. We human beings are unlike all the millions of other species that have ever evolved in that only we are aware of the very process of evolution. We can scan the history of this planet most of the way back to the origins of life itself. Our view of that history, in particular our understanding of humanity's place in it, is bound to color our attitude to the future and to affect whatever action we take (or fail to take).

As we see it, evolution took a sudden new turn with man's emergence, 2 to 3 million years ago, on the African plain. Until then the sole medium of change had been the DNA that formed the genetic material of all plants and animals. No change was secure until it was written into the DNA of a species—and all such changes occurred by accident, for no living thing could *will* a change in its DNA. But with the coming of

The world's have-nots (right) must spend so much time on the basic business of earning a bare subsistence that they have little chance or incentive to gather capital, although that is their only hope of economic improvement. Among them economic growth is naturally slow. At the other end of the scale are those whose every material want is satisfied (below). They are unlikely to strive as hard for their fourth and fifth car as people do for a first car. So among the world's richest nations effort slackens and economic growth is also fairly sluggish.

human intelligence, which can sustain complex language, writing, and mathematics, evolution no longer worked primarily through accident— not, at least, in our case. Today, certainly, we can bring about change through our consciousness— through looking, examining, trying to understand, and then taking certain actions as a result of that understanding. That is why it is so important for us to get our understanding right. We have a choice between a future in which we permit ourselves to slide into catastrophe and a future in which we can remain a viable part of the living world.

If, for example, we insist that economic growth leads to ecological disaster, and that the disaster is brought closer by each step forward of the world's poorer countries, it is like telling survivors of a shipwreck that there is room for only a few in the lifeboat and that, anyhow, the frail boat is sinking. It is, indeed, an open invitation to those already aboard the lifeboat (that is, the developed nations) to push the others into the sea. On the other hand, if we believe that quite another mechanism is at work, and that at a given level of prosperity a country's economic growth has a built-in tendency to slow down, thus making long-term stability possible, our whole attitude changes, and we discover that there is plenty of room in our big, tough boat.

Long-held beliefs in these areas are apt to be so firmly entrenched that they can even turn falsehood into truth. If a bleak faith in ecologic disaster and the big bad economic growth conquered enough minds, we could all become like people fighting for room in a sinking lifeboat. Under such circumstances, human life would be so unpleasant that all the gloomy predictions would no doubt be fulfilled.

To end on an entirely personal note (for how can one summarize *impartially* so many elegantly computed alternatives of hope and horror?), I believe that evolution, and especially the evolution of our own species, has more lessons for us than we have yet realized. Life began perhaps 3600 million years ago. For 2400 million years evolution did not get beyond the single-cell way of life. It took another 800 million years to evolve the first fishlike creature with a backbone. Every later development has been squeezed into the last 400 million years. Animals did not really master the land until 300 million years ago. Birds, mammals, and flowers became established only in the last 100 million years. Man has been

A NASA proposal for harnessing the sun's energy out in space and beaming it down to earth. The gigantic solar radiators will be designed to orbit in space some 22,500 miles away from earth. As in existing spacecraft, the giant solar-cell arrays convert solar energy into electricity, which is then transmitted back to earth by radio. Preliminary work on this project has already started.

around for only a couple of million years, and civilization for only 10,000-odd years. If all those events were compressed into just 12 hours, civilization would be less than a tenth of a second old. Our understanding of this fascinating process dates back just to the last century, and already we are capable of making intelligent use of all that we have learned.

There must have been many moments during our long journey from the African plains when people have thought, "Here we stand at a watershed in the tide of human affairs." So it may be rather presumptuous of me to stake out my own watershed claims. Still, I believe that there have been only three great moments of that kind so far, and that we are close to the fourth, which may also be the last.

The first really has to be the first. It is the time when the genus *Homo* emerged as a new element in the animal kingdom, long before the species *sapiens* came along and extinguished all his cousins. The original *Homo* was a nonhuman animal, cradled in the primeval forest. As with other nonhuman animals, reality for him was exclusively the here-and-now. Things that could be seen or heard, touched or felt, smelled or tasted—these were real, and they were all that was real. And *Homo*'s green, wordless world was the first world we had to navigate. The craft in which we made the journey, our animal body, is with us still. It bears all the marks of its origin, and when our guard is down it still gets up to its old tricks. Man's present bursts of passion, jealousy, rage, love, will to self-sacrifice, communal spirit, and ruthlessness are his clues to the long, wordless journey that his predecessors made through the green forest.

At length our ancestors stood on the far side of the wordless realm. They had at last reached an upland with strange new dimensions, lit by a fleeting, disturbing, contradictory light: the light of a language that enabled man to share ideas as well as food and shelter. It was the second great watershed in human affairs.

We may disagree as to the date of our first hesitant steps onto that new plane. It may have been 50, or 100, or 1000 millennia ago. But we can

This century has given us unique chances to see ourselves as a species apart from and unlike all others, especially in such giant gatherings as military parades and pop festivals. Now we must also learn to see ourselves as part of a world ecosystem for whose future we have a major responsibility.

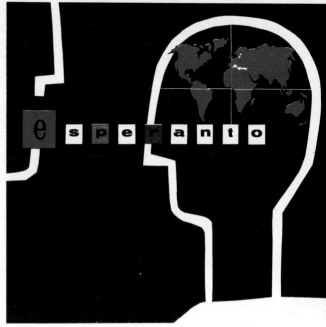

be sure of the result: an epidemic of words, a torrent of ideas. What was it like for those people to turn around and survey the rest of creation, and to realize that they were different? How did it feel to be able to communicate that awareness, to speculate about it? Did they understand that by saying "Do you think . . . ?" or "Is that possible . . . ?" or "Might this not be the case . . . ?" they were forever removing themselves from the primeval forest?

Words must have seemed at first to have magic power; words could apparently do everything. But the wordstorm that swept through their minds brought new dangers, without counterpart in the animal world behind them. For an awareness of ideas was bound to lead to an awareness of the mysteries, the unanswerable questions, that lurk in the world beyond the green and wordless jungle. The new world was a prison, after all—unless, perhaps, there was some spirit or essence that provided an answer to all the hard questions, a solution to the puzzles. And so humanity reached its third great watershed. First there had been the green animal world. Then came the magnificent word-magic world. Now our forebears entered what I like to think of as the Great Spirit world.

It was the Great Spirit world because He had created it. Indeed, it was not a world, it *was* Creation. And there at its apex, the earthly

Language, symbolized at far left by Claes Oldenburg's sculpture Alphabet/ Good Humor, *was a major landmark in our evolution from anthropoid to man. It is the medium by which we understand the world, share that understanding, and influence one another to belief and action. That influence can be enormous, as Lazarus Zamenhof realized when he invented Esperanto (left) as a potential world language in 1887. By contrast, Adolf Hitler (right) understood the power of language to reach into the darker side of man's soul. His speeches in the 1920s and 1930s welded the German-speaking peoples into one of the most destructive war machines in history. His infamous book-burnings (below) were aimed at quelling opposing argument—a black compliment to the immense power of words.*

The sculpture (extreme left), of fiberglass and bronze, 142″ × 68″ × 28″, Collection of Michael Crichton, Courtesy of Margo Leavin Gallery.

image of the Great Spirit, stood man. Different. Proud and transcendent. *Our* kind of man.

By the time we reached this great moment in our history, we already lived in cities, far from the forest world of hunters and nomads. Looking out over vistas of brick and mortar, it was easy to forget the old fears, easy to strip the rocks and trees—and even other animals—of their mystery and banish them to an earthly limbo. The Great Spirit, we knew, had created everything for us. This was a storehouse world, ripe for plunder. We could now set out to subdue the earth and have dominion over not only the fish of the sea and the fowl of the air, but over everything, whether animate or inanimate.

The noblest prophets in our history—Buddha, Christ, Mohammed, Confucius, and others—inferred from the Great Spirit a common humanity that transcended nations and even continents to unite all mankind in one commonwealth of dignity and love. If we had truly followed their teaching, we could have made the leap into a new way of life long ago. We were not ready to listen and leap, however. Totally divorced from the world in spirit, viewing it from the standpoint of privileged interlopers, we became increasingly detached. Our detachment was too casual to be considered brutal. What we felt about our planet was curiosity—but curiosity without concern. In time we even invented a new word for our privileged, inquisitive, casual detachment. We called it *science*. Science came to be an alternative insight to the insight granted by religion.

And so in our century we have once more discovered what it is like to be frightened by mystery—this time by the apparently unanswerable questions posed by our own very real technological powers. Coldly, calmly, 20th-century technological man has brought himself to a fourth watershed where there must be a great leap forward if he is to survive. It is no longer possible to walk coldly and calmly along the old familiar paths, for it is those paths that have led to an age in which we must contemplate possible futures in human megadeaths—or in uncountable, unimaginable numbers of deaths if we once again remember that we are only a small segment of earth's living beings. One of the futurologists' most terrifying scenarios *could* come to pass if we permit it to. If it does, it will have been the last lunatic play of a mind that divorced itself successively from earth, stones, rivers, and the rest of the living world—even, ultimately, from its own body, the battered vehicle that has brought us to the threshold of our fourth great moment in history.

We who live at the time of this fourth watershed may recognize that it is happening but may date the precise moment of its arrival differently. For me it came at the unforgettable hour when *Apollo*'s cameras swung aft, and there, hanging in the hostile void, still green and blue, lovely beyond words, was our tiny spaceship home, a mote of dust in an awesome emptiness. And it became, as it were, a diagram of all the worlds we have navigated since we quit the green jungle of our cradle. There, too, the mind's eye caught a glimpse of everything that science had put at risk—revealed to us in an instant by the latest and greatest of science's many miracles.

For me there is still something about that moment of insight that can choke the heart and mind. What I gain from it—and I know that many others must feel as I do, even if they would express it in quite different terms—is a better understanding of our remote forebears who felt and respected a Great Spirit in everything that clung to that blue-green island in the void. I feel sure that they had an intuition that we must above all seek to recapture—not in their crude and unsatisfying terms, to be sure, but in harmony with all those other intuitions that we have gained on the long journey since their time. Somehow we must learn to look at trees, and blades of grass, and beetles, and even pebbles and never say, never think, never feel that we have "dominion over" them. Rather, we must learn to think in terms of "sharing with," "uniting with." In doing so, we can unite our fragmented selves as well.

If we learn that lesson, we can still win out. The fourth world across which we must travel is within ourselves. The journey has already begun.

Or so we must hope.

We have seen the world from space, and space from the world. Now we know their history and their workings in such detail that we readily forget their essential one-ness—something our ancestors understood without difficulty. If we cannot regain the sense of unity in all things, the brute technology we have built for ourselves could easily destroy the living world.

Index

Picture Credits

Key to position of picture on page: (B) bottom, (C) center, (L) left, (R) right, (T) top; hence (BR) bottom right, (CL) center left, etc.

Cover: Photo Michael Freeman
Title page: *Things to Come*, London Films
Contents: Logica Ltd./Photo Clive Bubley
9   Photo Mike Busselle and Richard Hatswell © Aldus Books
10(T)   Drawing by Chas. Addams; © 1946, 1974, *The New Yorker Magazine*, Inc.
10(B)   *Dr. Strangelove: or How I Learned to Stop Worrying and Love the Bomb:* Dir. Stanley Kubrick, Columbia Productions
11   from *Tales of Wonder No. 4.* British edition published by Worlds Work Limited
12(L)   © Commander Gatti Expeditions
12(R)   Julian Calder/Susan Griggs Agency
13   Photo Mike Busselle and Richard Hatswell © Aldus Books
14   The Mansell Collection
18(T)   Hirmer Fotoarchiv, Munich
18(2nd)   Photo John Freeman © Aldus Books, courtesy of The Marquess of Salisbury
18(3rd)   Society for Cultural Relations with the USSR.
18(B)   Radio Times Hulton Picture Library
19(T)   The Mansell Collection
19(2nd & 3rd)   after Camera Press
19(B)   after *Daily Telegraph*
20   Anthony Frewin, *One Hundred Years of Science Fiction Illustration 1840–1940*, Jupiter Books, London, 1974
23   René Burri/John Hillelson Agency
27   after Marks, *The Dymaxion of Buckminster Fuller*, Reinhold Publishing Corporation, New York
32–3   after material by Professor Lawrence Klein
34   Kobal Collection
35   after Edward de Bono, *The Dog Exercising Machine,* Jonathan Cape Ltd., London
36   Burt Glinn/Magnum Photos
37   Bruno Barbey/John Hillelson Agency
39   Photo Michael Freeman
40–1   after Charles Jencks, *Architecture 2000*, Studio Vista, London
42   *Sunday Times Magazine*
42(B)   Photo Robert Estall
44   Mary Evans Picture Library
45   NASA
46   Anthony Frewin, *One Hundred Years of Science Fiction Illustration 1840–1940*, Jupiter Books, London, 1974
47   United Kingdom Atomic Energy Authority
48(TL)   Wolfgang Winter/Camera Press
48(B)   The Trustees of The Imperial War Museum, London
49(T)   Photo Mike Peters
51   Norman Myers/Bruce Coleman Ltd.
52(T)   Keystone
52–3(B)   after Charles Jencks, *Architecture 2000*, Studio Vista, London
54   Lee E. Battaglia/Colorific
55   Raghubir Singh/John Hillelson Agency
56   after Arnold Toynbee, *Cities of Destiny*, Thames and Hudson Ltd., London
58   Burt Glinn/Magnum Photos
59   Planned Parenthood Association
62   Aldus Archives
64   *Sunday Times Magazine*
65   John Garrett/Susan Griggs Agency
66   Mary Fisher/Colorific
67   Drawing by Chas. Addams; © 1969 *The New Yorker Magazine*, Inc.
69(TL)   The Mansell Collection
69(TR)   Peter Damian Grint
70(L)   Tom McHugh/Photo Researchers Inc.
71   Georg Gerster/John Hillelson Agency
73(T)   National Film Archive
73(B)   G. R. Roberts, Nelson, New Zealand
74(L)   Photo Michael Freeman
75   Günter Radtke
76–7   *Sunday Times Magazine*
78   Keystone
80(T)   Eve Arnold/Magnum Photos
81(T)   Dennis Stock/John Hillelson Agency
80(BL)   Photo S. Kingley-Jones © Aldus Books
81(B)   Tom McHugh/Photo Researchers Inc.
82   Günter Radtke
84   Don Hunstein/Colorific
85   Paul Fusco/Magnum Photos
87   *Rollerball,* Produced and Directed by Norman Jewison and distributed by United Artists
88–9   NASA
91   United Kingdom Atomic Energy Authority
92   Ultimate Publishing Company, New Jersey
94   Georg Gerster/John Hillelson Agency
95   Nigel Bonner, British Antarctic Survey, Cambridge, England
96–7   Popperfoto
98   Victor Englebert/Susan Griggs Agency
99(T)   Birmingham Museums and Art Gallery, England
99(BR)   David Hiser/Photo Researchers Inc.
100   W. Eugene Smith/John Hillelson Agency
102–3(T)   Leslie Brown/Ardea
103(B)   Russ Kinne/Photo Researchers Inc.
104   Edward P. Leahy/Camera Press
105(T)   Edwin Brooks/Camera Press
105(B)   Edward P. Leahy/Camera Press
110   Thomas Gilcrease Institute of American History and Art, Tulsa
111   Bob Moraczewski, *Big Farmer Magazine*
113   The Upjohn Company, Kalamazoo, Michigan
115   Photo Michael Freeman
116–7   Derek Bayes/Aspect
118   WHO photo
119   Anthony Frewin, *One Hundred Years of Science Fiction Illustration 1840–1940*, Jupiter Books, London, 1974
120–1   Photo Michael Turner © Aldus Books, by permission of Professor E. C. Zeeman, Professor of Mathematics and Director of the Mathematics Research Centre, University of Warwick, Coventry
122   after ed. L. Collins, *Use of Models in the Social Sciences,* Tavistock Publications Ltd., London, 1975
125–7   after Professor Dennis L. Meadows et al, *Limits to Growth,* Earth Island, London
129   Photo Michael Freeman
131(T)   Photo Robert Estall
131(B)   David Reed
133   Günter Radtke
134   Burt Glinn/Magnum Photos
135(R)   Photo Michael Freeman
136(L)   F. J. Thomas Photography, Los Angeles
136(R)   Universala Esperanto-Asocio.
137   Camera Press
139   Photo Michael Freeman

Artist Credits

© Aldus Books: Diagram 27, 32–3, 40–1, 43, 52–3(B), 56, 68, 93, 108–9, 122, 125–7; Ray Jellife 24–5, 30–1, 35; Vernon Mills 60, 63

GUIDE:
ECOLOGISTS
AT WORK

By Michael Hassell
and Stuart McNeill

One of the commonest methods of sampling insects on vegetation is to beat them from a branch onto a collecting tray.

Series Coordinator	Geoffrey Rogers
Series Art Director	Frank Fry
Art Editor	Roger Hyde
Design Consultant	Guenther Radtke
Editorial Consultant	Donald Berwick
Series Consultant	Malcolm Ross-Macdonald
Editor	Allyson Rodway
Copy Editor	Maureen Cartwright
Research	Enid Moore

Contents

Editorial Advisers

DAVID ATTENBOROUGH Naturalist and Broadcaster

MICHAEL BOORER, B.SC. Author, Lecturer, and Broadcaster

Introduction

The current concern that man's activities are endangering the delicate balance of plant and animal life in natural communities has made "ecology" a household word. In spite of this, there is little awareness of the breadth of work being done by ecologists throughout the world. Although some of this effort is directly concerned with man's impact on the environment, there are also many ecologists who are attempting to understand the basic laws governing the ways in which natural populations and communities function.

This book focuses on the work of just a few ecologists who have made important contributions to the whole subject. They have been selected to illustrate the different ways in which ecologists undertake their work. Some of the studies involve direct observation of natural populations and communities over many years. At the other extreme are examples of short-term laboratory experiments carried out under controlled conditions and designed to answer specific questions. They are all united, however, in having the same fundamental aim: to understand the laws of nature.

The Science of Ecology

Ecology is the study of the relationships of plants and animals to their environment. The concept of environment in this context is a very broad one, for it includes not only such physical factors as temperature, water, and the terrain, but the organic community as well. That is why ecologists have to study an extraordinarily wide range of problems, ranging from the interactions of living creatures of all kinds (competition, predation, and parasitism, for example) to the effects that temperature, the availability of water, and many other physical factors have on the survival and reproduction of a single species. Ecologists are all biologists, of course, but they belong to a branch of biology that directly seeks to understand natural communities. To do this they must study individuals, populations, and also the whole ecosystem within which the community operates.

Not surprisingly, in view of the breadth of interest that their discipline demands, ecologists must also draw upon the findings of many other disciplines both inside and outside biology. Foremost among these are the fields of taxonomy (the identification and classification of organisms) and physiology. An ecologist must be able to identify a species and its habitat with extreme precision in order to interpret the results of his study both to his own satisfaction and to that of his fellow scientists. Knowledge of physiology is important for anyone studying the responses of an individual organism to the physical environment. How does a given creature, a limpet for example, react to temperature, humidity, salinity, and so on? To know the answers to such questions is to hold the key to an understanding of why different species can survive only in certain limited environments. Thus, for example, an important factor in explaining the obvious bands, or *zonation*, of plants and animals on a rocky shore is their different tolerances to drying out during those periods when they are not actually covered by the sea.

Studies of behavior are also an integral part of much ecological work. Through them we learn about the many ways in which individual organisms have become adapted to the special ways of life necessary for them to survive in different environments. For instance, many desert-living animals survive only by occupying deep burrows during the scorching daylight hours and coming out for food in the cool of evening or night. Added to this they often have such physiological refinements as the ability to produce highly concentrated urine and to retain in their bodies much of the water that forms from the metabolism of food. Thus, from physiological and behavioral studies, the ecologist is better able to understand how some plants and animals can survive in the harshest environments.

Several branches of human knowledge that are entirely outside the limits of biology also play an important part in ecological research. Geography and geology, for example, are involved in understanding the distribution of plants and animals. The ecologist may also have to be a part-time sociologist and economist, because these disciplines are concerned with the effects of mankind on the environment. As ecology develops into a more rigorous science, mathematics is becoming increasingly important as a language of ecological theories.

Over the past 150 years, ecology has developed from being something of an art into a true scientific discipline. Learned men and women have always been interested in collecting and classifying the plants and animals around them, even if mainly for medical and culinary purposes. It is only when such activities are joined with an urge to examine and record the habitats of living organisms as well as the organisms themselves, however, that we have the beginnings of an ecological study. In fact, the ecologist is not always required to make actual collections of the specimens that interest him. He may base his research on acute observation alone or rely on the taxonomic skills of others. The extent to which the art of observing nature was carried in the days before ecology became an experimental and systematic science is exemplified by the work of some of the great 18th-century naturalists. Many of them lived and wrote in England and were among the first students of ecology in modern times. Perhaps the best-known of all was the Reverend Gilbert White, of the small village of Selborne in Hampshire. White spent countless hours observing the plants and animals of the surrounding countryside. He painstakingly noted down almost everything that he saw, and incorporated their descriptions in long letters to friends and acquaintances. A collection of the letters was published as *The Natural History and Antiquities of Selborne* in 1789. Even today, this delightful classic remains a source of inspiration to many biologists.

The evolution of ecology from something of an art into a scientific discipline (in which all ideas and hypotheses are rigorously tested) stems in part from the work of one of the greatest biologists, Charles Darwin. His epoch-making book *On the Origin of Species* (1859) consists largely of evidence in support of his theory that evolution by natural selection depends upon "the survival of the fittest"—a famous phrase originated by Darwin's contemporary Herbert Spencer, but approvingly quoted by Darwin himself. It is a phrase that embodies a wealth of basic ecological principles. Darwin observed that populations have a high potential for reproduction, but that this is counteracted by the effects of such processes as competition, predation, and disease, all of which adversely affect the less well-adapted individual organisms. This observation is rooted in the fundamental ecological fact that births and deaths within a population of animals or plants are in balance if population size remains constant.

The way that Darwin developed his theory reveals the approach of the true scientist. During an early career as a classical naturalist he traveled widely, collected many animals, plants, and fossils, and made a great number of significant biological and geological observations. Thus he slowly acquired the background knowledge that permitted him to develop his theory of evolution. In the following years he strove to validate these ideas by assembling a vast array of supporting evidence. This procedure is at the heart of the scientific method; on the basis of known facts, the scientist infers a general conclusion (or hypothesis), which he tests by unbiased observation and/or experiments. His further findings and those of his colleagues may support the hypothesis, invalidate it, or lead to its modification. Ecological theories are constantly being tested by observations and experiments both under natural conditions and in the laboratory, with one series of experiments often leading to the development of new theories that create a need for further tests—and so on.

As an illustration of this procedure let us set up a hypothesis of our own. Let us suggest that competition for food tends to become more severe as population density increases. We begin by choosing an appropriate experimental system, such as the competition for food by blowfly maggots. A standard amount of the food (ground liver) is placed in a simple container, and a few blowfly maggots added. The number of flies that later emerge is recorded. The next step is to repeat the experiment with a larger number of maggots, and then again and again, each time increasing the maggot population. The likely result is that the percentage of surviving flies decreases as the maggot population grows, because the limited amount of food causes many of the crowded population of maggots to starve to death. We might also find that those that do survive to adulthood are smaller in size than the blowflies that come from fully fed maggots. Eventually, as crowding increases, there will not be enough food to provide any of the maggots with sufficient nourishment to complete growth, and so not a single one will survive to adulthood. Thus, the experiment would confirm our original hypothesis that the intensity of competition depends on the supply of available food—at least in a controlled laboratory situation, using blowfly maggots as subjects. Obviously, further experiments are needed before we can fully understand competition in blowflies. We must answer such basic questions as: what are the effects of variation in size on the behavior and reproduction of the adult flies? The next stage is to ask more general questions: does widespread competition occur under natural conditions as well as in the laboratory; and to what extent does the hypothesis apply to other and very different species?

Like other scientists, ecologists need to be clear in thought and to have an enormous fund of patience and concentration. In the following chapters we take a look at some of the pioneering work that such people have done in several quite different areas of ecology. Taken as a whole, their achievements illustrate many of the basic principles underlying the science, but the findings were garnered in a variety of ways; some were the results of laboratory experiments, some involved the study of particular populations in nature, and some came from field studies of complex communities. This range of approaches indicates how broad a subject ecology is, and how many different skills must be harnessed in order to advance our understanding of the inter-relationships of plants, animals, and the environments in which they live.

Each of the ecologists in these pages has made a significant contribution to the development of ecology as a scientific discipline, but this is not to imply that there are not many others whose contributions are equally meaningful. Ecology has a significance for mankind that is far more profound than its simple definition suggests. To understand natural communities is to recognize our dependence on them. Man is a major component of many natural ecosystems, and the laws that govern all natural populations apply to him too. Thus, appreciation of the ecological principle that populations remain constant in number only when births are balanced by an equivalent number of deaths, has obvious relevance to man's current population explosion. Similarly, it is easy to speak glibly about conservation and the need to preserve wildlife. But only if we fully understand the basic principles of ecology can we manage our natural resources and the environment in general with the intelligent understanding that practical conservation requires.

Competition can occur whenever two or more individuals are striving against each other to secure a resource, such as food, that is in limited supply. The effects of competition therefore become more severe as the density of competitors increases. This is clearly illustrated by the simple experiment depicted here, where the density of blowfly maggots competing for a fixed amount of ground liver is varied. The outcome, in terms of the number of surviving flies, is shown in the bottom row of circles.

Competition for Food by Blowfly Maggots

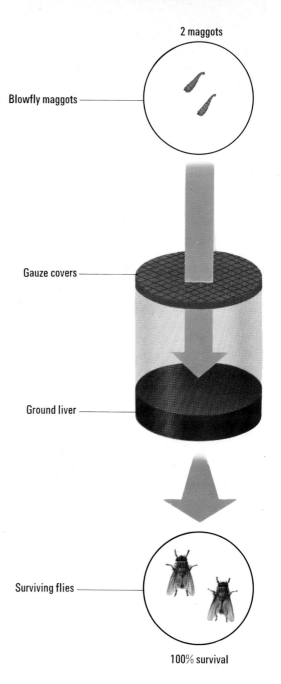

2 maggots

Blowfly maggots

Gauze covers

Ground liver

Surviving flies

100% survival

HYPOTHESIS : competition for limited food becomes more severe as population density increases

CONCLUSION : hypothesis confirmed for blowflies

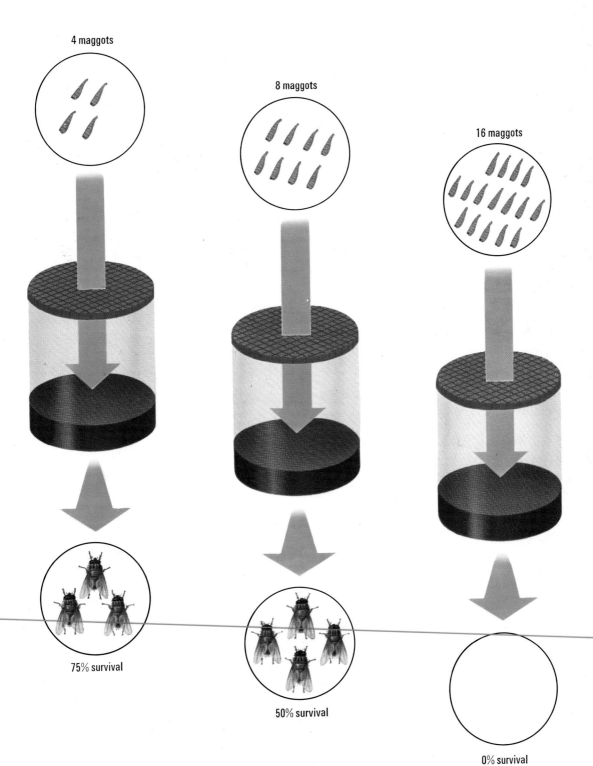

4 maggots

8 maggots

16 maggots

75% survival

50% survival

0% survival

Populations in Nature

Specific animals and plants vary greatly in their abundance. Some, such as the lady's-slipper orchid, are characteristically rare; others are very common; but most fall between the extremes. This range can be found within almost any natural community—from desert to marsh, from pond to ocean—and so the ecological principle that species have characteristic levels of abundance is easy to demonstrate. For example, all we need do in a typical meadow is to take a sample of its plant life by randomly dropping a 12-inch-square wooden or metal frame (called a "quadrat") and recording the numbers of each species within the squared-off area. After doing this many times, we shall inevitably have discovered that only a few species—several grasses, daisies, and buttercups, for instance—are common enough to appear over and over again in almost all quadrats. On the other hand, many species, such as orchids, cowslips, pasqueflowers, autumn crocuses, and so on, will have appeared quite rarely.

Even if we repeated this experiment over a number of years, we should find the general picture little altered. Although plant and animal populations do fluctuate measurably from year to year, it is usually over a very small range on either side of their average level. However, there are exceptions—for example the locusts, which periodically form immense migratory swarms that ravage considerable areas of the tropics and subtropics. Among several other species whose populations occasionally undergo spectacular fluctuations is the Canadian spruce budworm, which defoliates large tracts of forest—but only at intervals of several years. The lemmings of Northern Eurasia and Canada are also well known for their periodic increases, sometimes leading to mass migrations of these small mammals. But such dramatic exceptions do not alter the common pattern of average levels of populations for given species.

What the population ecologist wants to know is why do different species have different levels of abundance, and what causes the observed fluctuations? To find answers to these questions, the ecologist needs to identify and study the main

Animals can be sampled in a variety of ways to estimate the size of their populations. This picture shows two of the simplest ways of sampling insects on vegetation. On the left, a net is emptied of its catch after being swept through the plants on the woodland floor. On the right of the picture, insects are being beaten from a branch onto a large collecting tray.

causes of variation in the number of births and deaths within a given plant or animal population. Some variations, of course, are due to such climatic factors as temperature, humidity, rainfall, and salinity. For example, a very hard winter kills many small birds, and heavy rainfall can mean death to certain insects, such as aphids and other plant lice. But populations are also much affected by such biological factors as competition, predation, and disease. Plants are grazed by herbivores, animals are attacked by predators, and both plants and animals are subject to disease and the effects of competition.

Competition occurs whenever individuals (whether of the same or different species) strive after a shared resource that is in limited supply. This may be food, space, available mates, or the somewhat less definable commodity of social status. The less successful competitors have the lesser chance of survival or breeding. Thus, for instance, the barnacles that live on exposed rock surfaces on the seashore must compete for a limited amount of space. The acorn barnacle, for example, is faster growing than a related rock-dwelling species, *Chthamalus*, and therefore tends to displace *Chthamalus* on all except the highest regions of the shore, where the acorn barnacle cannot survive. And so we find the two species occupying separate zones on the seashore. Similarly, male fiddler crabs compete for space on the shore, but among themselves rather than with other species. Each male must secure a small area around his burrow in order to attract a mate; he usually does this without actually

fighting, by means of an elaborate claw-waving ritual that involves one much enlarged, brightly colored claw.

A number of ecologists have greatly broadened our understanding of the relative importance of such physical and biological factors by making long-term studies of single populations. Their work enables us to identify the major factors responsible for average levels of population abundance as well as the main causes of fluctuations. An important center for population ecology has been the University of Oxford, where research workers have carried out long-term field studies within a nearby piece of woodland called Wytham Wood. This area—which has become one of the most intensively studied areas of woodland in the world—includes two small hills, collectively known as Wytham Hill, that nestle within a loop of the River Thames to the north of Oxford. In its 1300 acres is a great variety of habitats, from open grassland with birch at the

13

top of Wytham Hill (which was the site of an ancient coral reef) to mixtures of such deciduous trees as oak, elm, ash, beech, and sycamore on the surrounding lower slopes. Ground cover within the wood varies from a predominance of dog's mercury on the higher ground to dense bramble and bracken lower down.

To discover how the student of population fluctuations does his work, let us focus on three of the best-known studies carried out at Wytham Wood, all of which began in the late 1940s and continued for many years thereafter. To begin with we shall examine the work on the winter moth carried out by Professor G. C. Varley and Mr. G. R. Gradwell of the Hope Department of Entomology at Oxford University.

The winter moth is a familiar pest throughout much of Europe. In addition to damaging fruit trees, it also feeds on a great variety of other trees, particularly oak. Occasionally, the caterpillars are sufficiently abundant to defoliate the trees entirely; and when that happened in Wytham Wood in the spring of 1948 and again in 1949, Professor Varley decided that this species was a suitable insect for a long-term population study, for it was not only abundant and easy to sample at different stages of its life cycle, but also a pest in some areas. An intensive study of the moth was obviously not feasible on a woodland-wide scale, and so Professor Varley chose to work within a small stand of 21 oak trees on the southwestern edge of the wood. Within this area he singled out five trees for continuous monitoring of their winter moth population.

In order to understand what the monitoring procedure involved, we need to know something about the life history of the insect. Winter moths, as their name implies, are active in the adult stage in the months of November and December. During the day, both sexes are inactive, but at night the females, which are wingless, crawl up the trunks of trees and release a scent, or *pheromone*, that attracts the males, some of which fly in from considerable distances. Mating takes place on the tree trunks, and the fertilized females then crawl farther up the trunk and along the branches to the terminal twigs, where they lay their eggs in small cracks in the bark and under lichen. A single female lays a total of about 200 eggs, after which she soon dies.

The tiny first-stage caterpillars emerge in April, and each tries to find an opening leaf bud in which to burrow and feed. If it cannot settle into a suitable nearby bud, it releases a silken thread that is caught by the wind, and the caterpillar may be carried a great distance. In early spring, therefore, a large "floating" winter moth population serves to disperse the species over wide areas. Of course, most of the wind-blown caterpillars fail to land on a suitable bud or leaf, and therefore die. The lucky few, however, feed rapidly. By middle to late May, having reached the end of their fifth and final larval stage, they begin to prepare for the long pupal stage during which they will metamorphose into adults. Initially, the caterpillars grow shorter and fatter, their bodies become opaque as fat is laid down, and their silk glands expand enormously. The silk has two uses: first, each caterpillar lowers itself on a silken thread from the crown of the tree to the ground below, where it quickly burrows into the soil surface. Then a silken cocoon is spun around the larva, and within this cocoon the pupa, or chrysalis, is formed. The adult moth emerges some five months later, thus completing its life cycle.

There are many hazards to be overcome at every stage of the cycle. Adult winter moths fall prey to spiders, beetles, mice, and other predators. Beetles, harvestmen, and various kinds of bug devour their eggs. Many of the hatching larvae never even manage to get their first meal, and those that are successful may be eaten by birds, or may be parasitized by creatures known as insect parasitoids. These are usually either small, wasplike insects or flies that lay their eggs on, in, or near some other insect, which the parasitoid larvae feed upon and eventually kill. Even the surviving caterpillars are far from safe, for many of them die during the long pupal period: some succumb to a disease of the silk glands, many others to predatory beetles, shrews, and insect parasitoids.

The task facing Professor Varley at the outset of his study was to find a dependable way to measure the numbers of winter moths on each of the five oak trees at successive stages in their life history. By doing this, he hoped he would be able to judge with some precision how many had died or disappeared between counts as a result of one or another of the causes of mortality listed above. The winter moth is a convenient species to sample at different stages because we have such a clear idea of where to find it. Professor Varley and Mr. Gradwell, who soon joined him in this endeavor, affixed traps, which worked on the lobster-pot principle, to the surface of the tree trunks. These traps were designed to intercept on average one quarter of all the females that were crawling up the trees during November and December. As the populations of males and females are about equal in size, this gave a measure of the total number of adults on each tree, and hence a rough idea of the number of eggs being laid.

Then, at the time of hatching, the two scientists placed sticky traps at intervals throughout the stand of oak within which they were working. These are small wooden trays mounted on stakes in the ground and covered in a substance to which any settling insect adheres. Because the trays cover a known fraction of the total area the numbers of tiny winter moth caterpillars caught on the trays give an indication of the numbers floating and settling in the whole area. Throughout the caterpillar period (April to June), Varley and Gradwell put weekly samples of twigs from each of the five oak trees into polythene bags and took them back to the laboratory, where the twigs were thoroughly searched for caterpillars, and the population of each twig was recorded. Next, in order to estimate the number of caterpillars dropping to the ground to pupate, 20-inch-square trays containing water in which trapped larvae would drown were placed under the five trees. The dead caterpillars were collected regularly, counted, and dissected. Dissection provided answers to two important questions: how many of the trapped larvae had been about to die from diseased silk glands? and how many of them contained parasitoid larvae and would have died soon?

The only other samples that Varley and Gradwell took were of adult parasitoids as they emerged from the underground winter moth pupae that they had consumed. For this operation they used the caterpillar trays again, but inverted and with small glass tubes let into the sides. Any insects coming up through the soil were attracted to this light source and trapped within the tubes, which were collected and replaced at regular intervals.

For more than 20 years they repeated this sampling routine, until at last they felt satisfied that they had a good understanding of how and why the winter moth population of Wytham Wood fluctuated from year to year. They found an approximately 35-fold variation, from as few as about 400 larvae per tree to as many as nearly 16,000, with an average abundance over the 20-year period of about 1500 per tree. From a detailed analysis of the sampling data, they calculated the numbers of winter moths that had died at every stage of the life cycle in each year and were able to identify three main causes of death: (1) predation of eggs and the failure of small larvae to find food; (2) predation of pupae by beetles and shrews; and (3) parasitism of both larvae and pupae by a fly, *Cyzenis*, and a small ichneumon wasp, *Cratichneumon*. Of these, (1) was the largest and the most variable source of mortality and it was thought that the egg predation was relatively unimportant compared with the limited ability of small larvae to find a suitable bud or leaf in which to burrow and on which to feed. Varley and Gradwell called this mortality the *key factor* because they found that its variations were largely responsible for the yearly population fluctuations that they had observed.

Their careful research indicated that the number of surviving caterpillars in any year depended largely on whether or not the eggs hatched at a time when the oak buds were beginning to open up and provide a food source. The winter moth as a species is in a critical position. If the eggs hatch too early, widespread starvation occurs; if at just the right time, the larvae flourish; if much later, however, they are again in trouble, for the older leaves do not provide a suitable food supply. This is because in June the leaves start to accumulate tannins and other compounds, and these chemicals inhibit caterpillar feeding and growth. So the time of hatching is of vital importance.

The pupal predators—mainly beetles—are also an important cause of mortality, and one upon which the weather has little effect. The number of winter moth pupae that the predators consume in any one year depends chiefly on the number of pupae available. Varley and Gradwell found that as many as 97 per cent were eaten in plentiful years, and as few as about 50 per cent in years when winter moths were scarce. Ecologists speak of a mortality that varies in this way as *density-dependent*, because the proportion of prey eaten by predators rises as the population density of the prey increases. This mortality has an important influence on the population changes of the winter moth. The key factor, remember, depended on whether or not the newly hatched larvae could immediately find food. In a very good year,

15

NOVEMBER-DECEMBER

JULY-SEPTEMBER

MAY

16

APRIL

Long-Term Study of the Winter Moth in Wytham Wood

1 Winter moth eggs on twig
2 Female winter moth trap
3 Female winter moth
4 Male winter moth
5 Shrew eating winter moth
6 Spider eating winter moth
7 Newly hatched winter moth caterpillar
8 Feeding 1st stage caterpillar
9 Floating caterpillar
10 Sticky trap for caterpillars
11 *Cyzenis* fly trap
12 *Cyzenis* eggs on leaf
13 Winter moth 5th stage caterpillar
14 *Cyzenis* fly
15 Great tit
16 Caterpillar spinning down from tree
17 Caterpillar tray
18 *Cratichneumon* wasp
19 Shrew
20 *Cratichneumon* wasp trap
21 Winter moth pupa
22 *Cyzenis* pupa
23 Carabid beetle
24 Staphylinid beetle

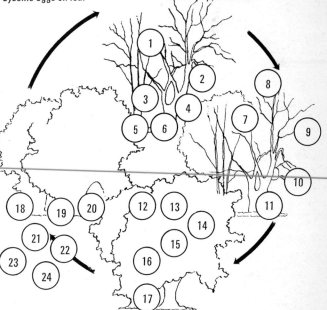

The winter moth has a 12-month life cycle. The wingless female moths emerge from the ground in early winter and climb the trees. Mating takes place on the trunks (photograph top right) and the eggs are laid on the outermost branches. The tiny caterpillars emerge in spring, and by early summer are fully fed and ready to drop to the ground to pupate just below the surface. The diagram shows some of the important predators and parasitoids of the winter moth as well as the traps used to sample the populations throughout the year.

The various kinds of winter moth traps catch a known proportion of the population at different stages of its life cycle. Left: the traps fixed to the tree trunks collect the female moths as they ascend to lay their eggs. Above: the trays containing water intercept a proportion of the fully grown caterpillars falling from the canopy to pupate below the soil surface.

when the key factor is small, the population would have been incredibly large except for the fact that a much greater than average percentage of the pupae were eaten by beetles and other predators. Thus, the density-dependent mortality factor kept the range of fluctuations in numbers from broadening too widely.

The effect of the parasitoids was less clear-cut. The researchers found that the parasitoid fly *Cyzenis* made only limited inroads on the population. The ichneumon wasp *Cratichneumon*, on the other hand, was responsible for the death of many pupae, and contributed significantly to the annual fluctuations.

The picture that emerges from the observations of Varley and Gradwell is a portrait of balancing factors, some of them causing population fluctuations, others tending to maintain the population at its average (or "equilibrium") level. An essential feature of populations that are in balance (that is, not consistently increasing or decreasing) is that, on average, there is only one member of a new generation to replace each adult that

dies. In the case of the winter moth, which produces some 200 eggs per female, this requires that an average of 198 of the offspring of each female must fail to complete their life cycle.

The Varley-Gradwell study shed new light on the ways in which this degree of mortality occurs. Its revelation of the different long-term effects of various mortality factors upon winter moth populations has done much to assist our understanding of such natural populations. Every such study, whether of animal populations or of plants, helps us toward a clearer comprehension of the ecology of our living world.

In 1947, a couple of years before the start of the winter moth studies, Dr. David Lack and other scientists of the Edward Grey Institute for Field Ornithology at Oxford University had begun a population study in Wytham Wood of a common woodland bird, the great titmouse (commonly called the "great tit"). This long-term study is in fact still going on. The great tit, the largest tit that lives in both the deciduous and the coniferous woods of Britain, breeds in tree-holes or in

The aspirator, or "pooter," is used a great deal by ecologists studying insects and other small invertebrates in the field. When the organisms have been collected in nets or on beating trays they can be sucked into the collecting tube of the pooter (above), from which they may later be transferred to another container. Below: a sample of insects in a collecting tube.

such artificial nest sites as boxes and drainpipes. It is a familiar sight in British gardens, where it can be seen feeding at bird tables, showing a marked preference for fat and nuts. Under natural conditions, though, it feeds on a wide variety of insects, spiders, and other invertebrates, as well as on seeds such as beechnuts when they are available during the winter months. In the spring, adult great tits nourish their young exclusively on insects and spiders.

This range of food was recorded by means of a 16-millimeter camera mounted in the back of special nest boxes. Whenever a parent bird returned to the box, it triggered the camera to take a photograph of its beak. In all, the scientists obtained about 29,000 photographs during this phase of their project, and the items of food held in the birds' beaks were identified by comparing them with collections of insects at Oxford's Hope Department of Entomology. The results showed that the diet of nestling great tits consists of a wide variety of moth caterpillars and pupae, together with other insects and spiders. The precise diet at any time depends on the relative availability of prey. Thus, for instance, winter moth caterpillars are an important food during the latter half of May, when, before pupating in the soil, they are generally abundant.

The great tit study area in Wytham comprised 63 acres of woodland. Within that area, the researchers began by attaching 200 nest boxes to tree trunks—such an overprovision of nesting sites that virtually every great tit within the 63 acres was likely to use them. This "saturation" technique is especially convenient for a population study because it allows easy counting of the entire population during the breeding season. Every breeding season since the study began, the number of great tits has been estimated at each of four different stages. First, Dr. Lack and his associates establish the number of birds breeding. This is done quite simply by seeing which nest boxes are occupied by great tits. Secondly, they determine the clutch size, or number of eggs laid by each female. This can be quite variable—anything up to 16. Then they count the number of chicks that hatch, and a comparison of this with the second figure shows how many eggs have failed to survive. Finally, they determine the loss of chicks by counting the fledglings that leave the nest. In each successive year, this spring count of breeding birds tells the scientists how many of the fledglings and their parents have survived the winter.

The first such count was made in the spring of 1947, after an exceptionally severe winter, when the breeding great tit population within the nest boxes was down to only 7 pairs. By the spring of 1948, the population had grown to 21 pairs, and by 1949 to 30. A gradual increase thereafter brought the total up to an average level of 45 pairs, with only small-scale yearly fluctuations until 1961, when the population rose dramatically, for some unknown reason, to 86 pairs. After that spectacular rise, however, it dropped back to the average level in 1962, where it has remained since. Once again, then, it is the average levels of abundance and the yearly fluctuations that are truly significant.

Their sampling routine allowed Dr. Lack and his colleagues to focus on four factors that evidently affect these levels and fluctuations: variations in clutch size, mortality of eggs, mortality of chicks, and mortality during the rest of the year. Some of the reasons for fluctuations are self-evident. Variations in clutch size are obviously related to the availability of insect food in the springtime, and also to the size of the breeding population, because a high population leads to more battles among the birds for territory and less time for collecting food (an important factor in that it is the male's responsibility to provide the female with much of her food as well as to defend the territory). The destruction of eggs is due in great part to weasel predation; in one year, the observers found that weasels had destroyed over 30 per cent of all the eggs laid. Much of the chick mortality comes from the same source. Most of the rest of the eggs that fail to hatch die because the parents abandon their nests, because of either human interference or the onset of sudden spells of cold, wet weather. In addition, from 5 to 10 per cent of the eggs are usually infertile.

But the cause of by far the largest mortality among the great tits of Wytham Wood is not fully understood. Many of them simply disappear between the time of leaving their nests as fledglings and the following spring. As their winter diet consists mostly of beechnuts, it seems probable that their survival rate is partly linked to the size of the beechnut crop. It is also likely that, as with other small-bird species of the temperate zone, severe winters can be a cause of widespread deaths. Some movement of the birds to and from the study area is also probable.

The long-term population studies on the tawny owl, great tit, and winter moth in Wytham Wood revealed a large number of interactions with other species. Here we see some of those interactions in a highly simplified food web. The tawny owl is a top predator in Wytham Wood and has no natural enemies. It feeds on several species of small mammals, especially mice and voles, and also on some insects. The great tit, on the other hand, is lower in the food web. Its main predator is the weasel, which often destroys many eggs and chicks. The great tit's diet consists mainly of small invertebrates and, in the winter, beechnuts. Near the bottom of the food web comes the winter moth, which is eaten by many predators and parasitoids.

These are only three, however, of several possible explanations. And so, just as the Varley-Gradwell study of winter moths found a key source of mortality in the inability of larvae to find suitable food, the study of great tits suggests a key cause of population fluctuations among the birds: the disappearance of many of them between the time of fledging and the next breeding season. The number of disappearances varies from year to year—it has been as low as 50 per cent and as high as 80 per cent.

The fact that the Wytham great tit population remains fairly close to its average level of 45 breeding pairs indicates that some other factors must be important in regulating the population. Among these, weasel predation is probably the most important; and, like beetle predation among the winter moths, it is density-dependent—in other words it is most severe when the population is most numerous. Significantly, variations in the average number of eggs laid per female is also density-dependent; that is, the smallest clutch sizes have been recorded in years when the birds themselves have been most numerous, the largest when there have been fewest pairs of breeders. So the great tit provides a good example of a species whose birthrate, as well as its mortality rate, has a noticeable effect on population regulation.

In Wytham Wood, 1947 also saw the start of another 20-year-long population study, on the tawny owl, by Dr. H. N. Southern of the Bureau

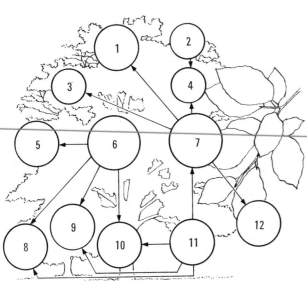

A Simple Food Web for Wytham Wood

1 Spider
2 *Cyzenis* fly
3 Green tortrix caterpillar
4 Winter moth caterpillar
5 Cockchafer beetle
6 Tawny owl
7 Great tit
8 Rabbit
9 Bank vole
10 Wood mouse
11 Weasel
12 Beechnut

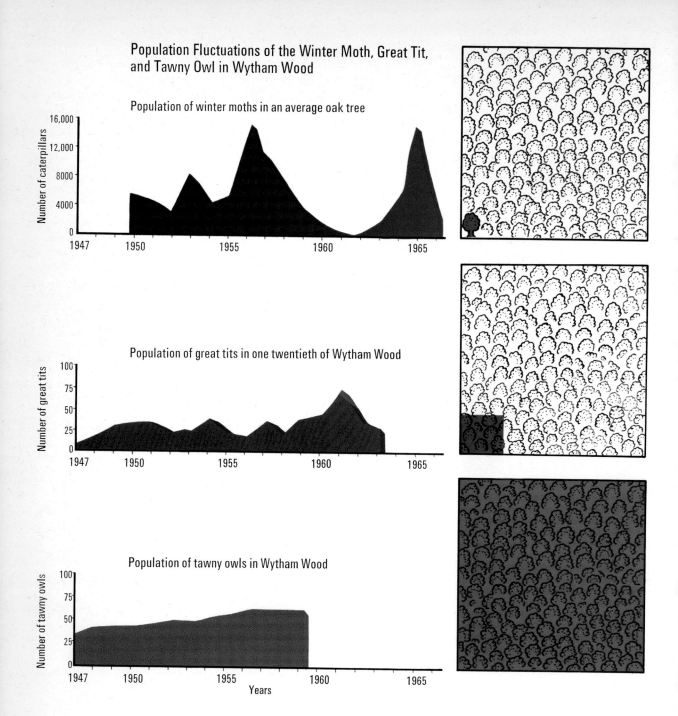

Population Fluctuations of the Winter Moth, Great Tit, and Tawny Owl in Wytham Wood

Population of winter moths in an average oak tree

Population of great tits in one twentieth of Wytham Wood

Population of tawny owls in Wytham Wood

Years

of Animal Populations at Oxford University. The tawny owl, the largest of British owls, is Wytham's chief bird predator. Primarily a woodland creature, it can also live in open country, as long as there are at least some trees; and its main source of food is mice and voles, although it will occasionally eat earthworms and large insects. Its usual hunting method is to drop silently on its prey from a vantage point in a tree. Like the great tit, it is territorial, but its

territory is a feeding one, and so the owl defends it all year round rather than solely during the breeding season.

Dr. Southern's study area was the whole 1300 acres of woodland, as against 63 acres for the great tit and 5 oak trees for the winter moth. These differences reflect the different-sized feeding areas required by large and small animals. Yet, even though such a large area was involved, Dr. Southern managed to count regularly the entire owl population. Because the owl is a nocturnal bird he had to do his counting at night, but the fortunate fact that owls are noisy as well as nocturnal enabled him to pinpoint each bird within its territory. He was also able to induce about half the owl population to breed in special nest boxes that he positioned in the trees. This permitted him to study breeding success in some detail, much as the breeding habits of the great tit were studied. Fledgling owls remain for some months within their parent's territory, and this too proved helpful for making population estimates.

As with most birds of prey, owls regularly regurgitate the indigestible parts of their food in the form of large pellets. Dr. Southern easily found many of these, and from them he soon discovered the birds' diet. Tawny owl pellets are largely made up of the fur and bones of mice and voles, together with some insect remains. During the summer, when the young are growing rapidly, the diet is broadened to include young rabbits, moles, and a variety of insects. Dr. Southern's analysis of the pellets also revealed that in winter the owls are remarkably dependent on just two species of prey, the wood mouse and the bank vole, without which the owl population is likely to starve. Indeed, after the hard winter of 1946–7, when mice and voles must have been in very short supply, the owl population was reduced to 17 pairs. Their numbers gradually rose to 30 pairs by 1955 and remained at about that level for the remaining years of Dr. Southern's long-term study.

Clearly, the availability of mice and voles is a critical factor for owl survival, and so Dr. Southern extended his study to include the populations of these vital food sources. His technique for sampling the small mammals was different from any so far described.

Every two months, he laid down 468 traps designed to catch, but not to harm, small mammals. Each trap consisted essentially of a 7-inch-long box containing food (oats) and bedding material (hay), with a trapdoor that closed behind the rodent when it had entered. All mice and voles captured were recorded, had a num-bered metal ring attached to a leg, and were then released. The size of the mouse and vole populations could be estimated from the proportion of ringed animals that were recaptured at a later date. By this method, Dr. Southern found that both mice and voles are most abundant in Wytham Wood in midwinter and fewest in early summer, but that there are times when they are scarce throughout the year.

The success or failure of the springtime breeding season for the tawny owls is related to the availability of mice and voles in the previous winter. In 1958, for example, when small mammals were exceedingly scarce, all the owls in the wood failed to breed, whereas in a normal year about three quarters of the owl population breed successfully. For the birds that do breed, low levels of food supply are also implicated in the failure of some eggs to hatch (because the female deserts her eggs to search for food), and in the failure of chicks to fledge or else to survive until the following spring. From all these details, Dr. Southern eventually concluded that such year-to-year variations as there are in the size of owl populations result from the relative success or failure of the breeding season, but that any changes in the failure of many fledglings to survive their first winter is responsible for the steady population.

The striking difference between Dr. Southern's findings and the studies of the great titmouse and the winter moth is that food, rather than climate or predation, is certainly the main limiting factor throughout the life cycle of the tawny owl. This fact, of course, reflects the owl's position in Wytham, where it has no natural enemies and is at the top of the food chain. The winter moth, on the other hand, is much nearer to the bottom of the food chain and is considerably affected by the weather as well as by various predators. The great tit, in an intermediate position between moth and owl, is less dependent on climate than is the moth, has only one major predator, and is sometimes food-limited.

The three projects discussed in this chapter are alike in that the research workers have chosen a particular population to study in considerable detail. From their results, they have discovered in each case the major factors affecting changes in the size of the populations. From the standpoint of ecology, such studies are important in showing us the kinds of factors that cause populations to fluctuate and to be rare or abundant.

Predators and Prey

Predators are an integral part of all animal communities, and often play a significant role in determining the numbers and distribution of their prey. The study of predation has therefore been a central one in ecology. Ecologists have observed and analyzed predation both under natural conditions and under the more strictly controlled conditions of the laboratory, and their findings have greatly broadened our understanding of the role of predators in nature.

All predators have one characteristic in common: they capture and eat other animals. In other respects, however, they can be very different. They range in size from the smallest unicellular organisms to the largest of animals, the blue whale, which eats the tiny shrimps known as krill. Predators are found in almost all habitats, from the poles to the equator and from the deepest oceans to the highest mountains. Although they are for the most part members of the animal kingdom, there are a few specialized predatory plants, such as Venus's-flytraps and pitcher plants.

True parasites, such as tapeworms and flukes, differ from predators in that they usually live on or in their hosts and are nourished by them without killing them. Inevitably, there are examples that fall between the two extremes of predator and parasite. Some parasites kill their hosts, as shown by many disease organisms, and there are predators that graze on their prey without killing them. Young plaice, for example, often feed on the siphons of some bivalve mollusks, which are able to regenerate their siphons later. A rather different category of predator is the parasitoid, which we met in the preceding chapter. Parasitoids are sometimes spoken of as "insect parasites"; mainly flies and wasps, they attack almost exclusively other insects, usually in the egg, larval, or pupal stage, and they differ from more conventional predators in that only the adult female searches for prey. Instead of eating her victim, she lays one or more eggs on, in, or near it, and the emerging parasitoid larva does the feeding. The insect prey is finally killed only when the parasitoid is fully fed and about to pupate. This is by no means an obscure group.

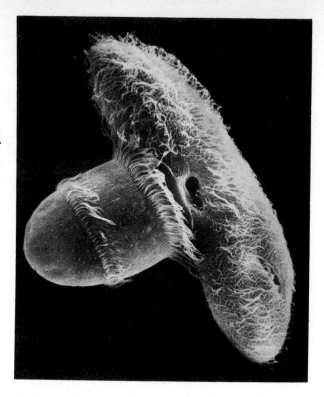

This series of scanning electron micrographs shows, from left to right, stages in the capture and ingestion of the slipper-shaped ciliate Paramecium *by the predatory ciliate* Didinium. *Simple animals such as these protozoans (shown highly magnified) make ideal subjects for studies in the laboratory, where the experimenter can observe them under controlled conditions.*

In fact, about one out of six of all insect species belongs to it—which means that roughly one in 10 of all animals from the protozoans to man is a parasitoid.

One reason why so many ecologists have been preoccupied during much of this century with the study of predation is that predators play such an important role in the control of other populations. Few species are free from the effects of predation; indeed, only those that are very large, such as elephants and whales, or are themselves predators at the top of a food web, such as lions, some sharks, and man, are likely to remain unscathed.

Early workers in pest control also appreciated the importance of predators—in particular, the potential value of predatory insects for the control of pest populations. This has led to the practice of biological control, in which parasitoids or other insect predators are introduced into an area to combat a particular pest. A classic example of biological control dates back to

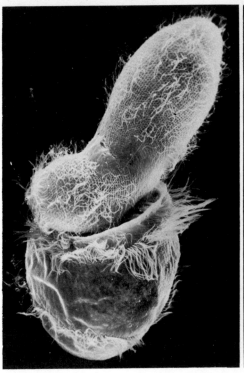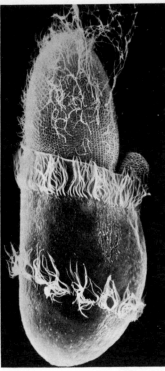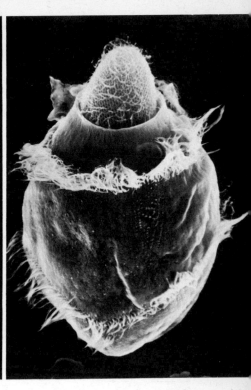

1889, when the vedalia beetle (a type of ladybug) from Australia was introduced into California as a predator of an insect, the cottony-cushion scale, which was threatening the future of California's citrus industry. The beetles soon reduced the scale populations to an average level so low that they caused no further economic damage. This stable situation, with pest and predator both rare, has persisted up to the present time, except in certain areas where insecticides directed at other citrus-fruit pests have severely reduced the vedalia population, giving the scales a chance to build up their numbers locally.

Some predators are themselves pests. Ecologists have been keeping a wary eye, for example, on the current epidemic of crown-of-thorns starfish in the western Pacific. These starfish are extremely effective predators of corals, and their populations have increased enormously in recent years. As a result, considerable parts of the Great Barrier Reef and other coral reefs have been destroyed.

The ways in which modern science approaches the study of predation are many and varied, but there are, basically, two major choices: the ecologist can look at predators and prey in their natural surroundings, or he can observe them under the controlled conditions of the labora-

tory. An example of the first of these alternatives is the much-studied interaction of the wolves and moose on Isle Royale, a small island in the Michigan sector of Lake Superior. Moose were first seen on the island early in this century; they probably got there by crossing from the mainland on winter ice. Their population rapidly increased to between 1000 and 3000 in the 1930s, and at such densities they were overgrazing their limited food supply. Widespread starvation followed, and the size of Isle Royale's moose herd was considerably reduced. As soon as the vegetation recovered, however, their numbers grew swiftly, and the sequence repeated itself.

This unstable situation, which ecologists were now observing closely, changed during the very harsh winter of 1949, when hungry timber wolves crossed the ice and became established on the island. The wolves fed almost exclusively on the moose and reduced their population to between 600 and 1000—a number for which the food supply was adequate. The wolf population was now maintained at between 20 and 25 individuals, and they were killing a sufficient number of moose to prevent them from again becoming too numerous for their food supply.

This example from the natural world provides evidence that predators can keep the numbers of their prey down to a steady or equilibrium

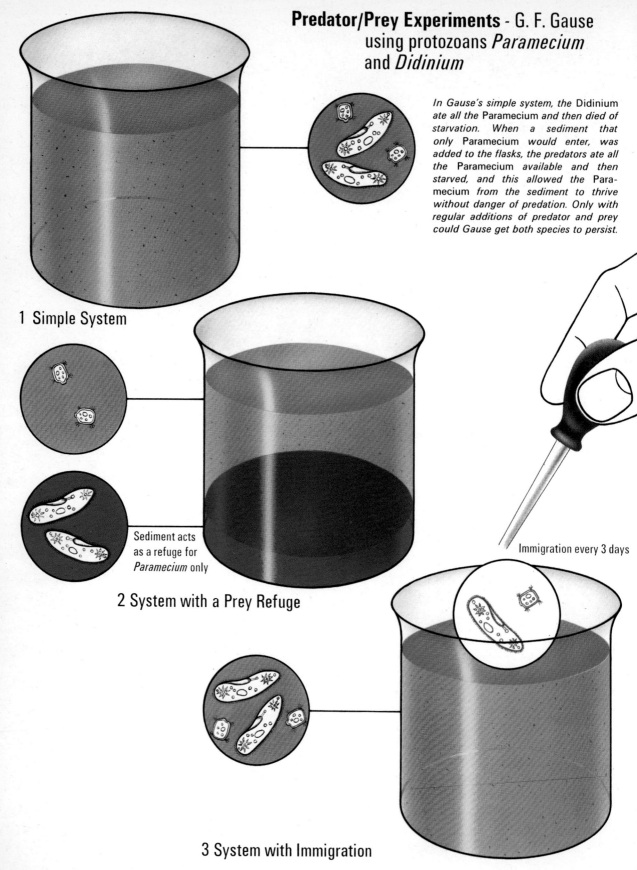

Predator/Prey Experiments - G. F. Gause
using protozoans *Paramecium* and *Didinium*

In Gause's simple system, the Didinium ate all the Paramecium and then died of starvation. When a sediment that only Paramecium would enter, was added to the flasks, the predators ate all the Paramecium available and then starved, and this allowed the Paramecium from the sediment to thrive without danger of predation. Only with regular additions of predator and prey could Gause get both species to persist.

1 Simple System

Sediment acts as a refuge for *Paramecium* only

2 System with a Prey Refuge

Immigration every 3 days

3 System with Immigration

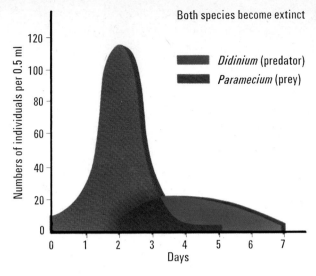

Both species become extinct

Didinium (predator)
Paramecium (prey)

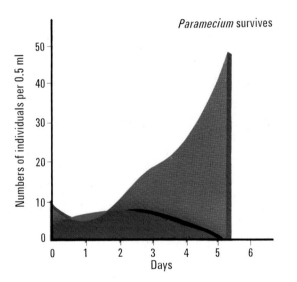

Paramecium survives

level. Recent reports suggest, however, that the moose population of Isle Royale is increasing slowly again, and ecologists are in some doubt as to whether the current increase will eventually be reversed by overgrazing, by an increase in the size of the wolf population, or by other factors. The problem with understanding the workings of a predator-prey system under natural conditions is that many external factors, such as disease, climate, and other types of predator, affect both the populations that are being observed. That is why some scientists have chosen the alternative approach of looking at the interactions of predator and prey in the simpler setting of the laboratory, where, with most variables under the experimenter's control, he can begin to understand *how* predators affect the population of their prey. Two famous experimental studies will serve to illustrate this.

The first is the work of the eminent Russian ecologist G. F. Gause, who made a study in the 1930s of the interaction of two species of microscopic protozoan. One of these, the familiar slipper-shaped *Paramecium*, is preyed upon by a smaller, related species, *Didinium*. It was Gause's aim to test the validity of a classic theory of the way predators and prey interact, which had been developed independently in the 1920s by an American mathematician, A. J. Lotka, and an Italian biologist, V. Volterra. According to this theory, there should be predictable oscillations of predator and prey populations—that is, the numbers should increase and decrease in a regular cyclic way. Such oscillations would seem to be

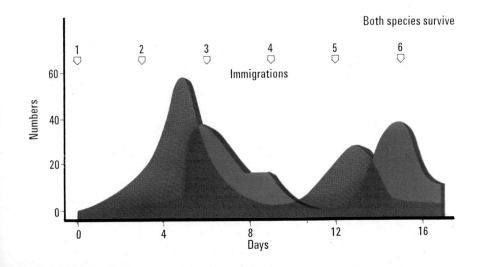

Both species survive

Immigrations

likely, because the prey diminishes as the predators eat their fill, and then, as food becomes scarce, the predators begin to starve and the prey can multiply again. Thus, the predator population cycles should lag just behind those of the prey.

Gause devised an extremely simple system of small glass tubes to serve as containers for his experimental populations. In each tube he put a filtered extract of oatmeal upon which a variety of bacteria were feeding. He then introduced into each tube *Paramecium*, whose sole source of food was the bacteria. On the next day, Gause added a few *Didinium*, and thereafter he examined daily samples of water under the microscope and counted the numbers of predators and prey. In this way he was able to estimate population densities as they changed throughout the experiment. He repeated the experiment again and again, and always obtained the same result, which contradicted the Lotka-Volterra theory. The *Paramecium* increased rapidly until the introduction of *Didinium*, which, being an efficient predator, soon caused the decline and extinction of its prey. *Didinium*, now without food, quickly suffered the same fate.

Gause then tried a second experiment. To create a secure refuge for the prey, he added to the oatmeal some solid matter that formed a cloudy sediment at the bottom of the tube. The *Paramecium* would readily enter the sediment but the *Didinium* remained only in the clear water. This slight change in experimental procedure completely altered the outcome. The *Didinium* ate all the *Paramecium* in the clear water and then starved to death. The *Paramecium* that were within the refuge multiplied rapidly to the limits of their bacterial food supply and spread throughout the now predator-free water.

So far, then, Gause had failed to find any trace of regular oscillations in the numbers of predators and prey. Now he proceeded to a third series of experiments. This time, he provided no refuge for the *Paramecium*, but on every third day he added one *Paramecium* and one *Didinium* to each tube. This affected the outcome dramatically, resulting in periodic oscillations in both species with those of the prey population preceding those of the predators—just as anticipated by Lotka and Volterra. Gause concluded that predator-prey interactions are inherently unstable, as shown by his first two experiments, and that repeated cycles in predator and prey numbers must depend upon external factors, such as immigration, rather than being a fundamental property of the interaction.

This conclusion was questioned in 1958 by C. B. Huffaker, an eminent American ecologist at the University of California, Berkeley. From a different experimental study, he was able to show that under certain conditions predators and prey can coexist, even in a laboratory system, without outside manipulation. He thought that Gause's failure to show this hinged upon an insufficiently complex experimental system.

Huffaker worked with two species of mite (tiny arthropods related to spiders). One species (the prey) fed on oranges and served as food for the predatory species. Huffaker began his series of experiments by setting up relatively simple systems. In one of these he used oranges partly covered with paraffin wax, through which the mites could not feed. The surface area exposed varied with the number of oranges so that the total surface area available to the mites remained constant. The oranges were placed in trays that had spaces for 40 oranges, but in most experiments less than 40 were used and the remaining spaces were filled with rubber balls, so that the total surface over which the mites could move was kept constant. In the first experiments the prey were introduced on their own. The numbers of mites were estimated at regular intervals by counting the number on a standard area of orange. The general result was that they increased rapidly at first and then fluctuated about an average level determined by the amount of exposed orange.

Huffaker followed these experiments with similar ones, but this time the predator species was also introduced. The outcome was now the same as that of Gause's first experiment: a single oscillation, followed by the extinction of first the prey and then the predator.

Huffaker now increased the complexity of his system. By combining three trays, he was able to use 120 oranges, each with only one twentieth of its area exposed to the mites. The rubber balls were not used in this system, but Huffaker placed barriers made of wood coated with sticky vaseline among the oranges. These impeded the easy dispersal of mites from orange to orange. On the other hand, to aid dispersal, Huffaker put small upright sticks at intervals in the tray, from which the mites could launch themselves on silken strands carried by air currents. Into this rather complex world, equal numbers of predators and prey were introduced. The result was a splendid example of just the type of continuing predator-prey population oscillations that Gause had been unable to obtain without the help of immigration.

The message is clear. Predation tends to produce cyclic changes in numbers of both predator and prey populations. In the simplest of natural systems, the changes may well be so violent that one or both populations will rapidly die out. But as the system becomes increasingly complex, there is a great tendency for both predator and prey to remain in a kind of oscillating balance.

Recent studies have revealed some very important components of predation, all of which contribute to the complex picture of predator-prey interactions. Consider what may happen as a ladybug larva searches for aphids on the leaf of a plant. The larva moves over the leaf at random, without reference to the presence of aphids, which it can detect only on direct contact. Once it encounters one, however, its powerful mandibles seize and devour the aphid. If it is still hungry, the larva resumes its search, but with a very much altered pattern, for it now carries on an intensive search in the immediate neighborhood of the spot where it has just eaten. This behavior is advantageous to the larva because prey generally tend to occur in clumps rather than singly, and the discovery of one victim often leads to another. If the larva has no quick success in the area, it gradually reverts to the more random kind of movement.

When, as frequently happens, the ladybug encounters another searching larva instead of an aphid, it is likely to react by moving to another leaf or even by dropping off the plant. This normal interference behavior has the ecological advantage of spreading predators over a large area and preventing local overexploitation of prey. So the number of aphids that the ladybug larva eats in one day depends on many factors. It must depend, of course, on the availability of prey, but it also hinges upon the area effectively searched by the larva, which in turn depends upon such matters as the speed of the larva's movement on the leaf, the time it takes to eat an aphid, the level of its hunger, and the frequency with which it bumps into other ladybugs. All these depend at least partly on the size of the

predator. A large ladybug larva moves faster, eats more quickly, and has a greater appetite than one that has just hatched.

All such kinds of behavior have been explored in laboratory experiments and shown to be common to a wide range of predators. Theoretical studies that include these complexities show that predator-prey cycles can still be obtained, but the tendency is toward greater stability of the populations as seen in Huffaker's experimental results compared with the violent fluctuations and extinctions of Gause's simple systems.

Regular large-scale cycles of population change are rare under natural conditions; and wherever they do occur, it seems unlikely that predation is responsible for them. The well-known regular fluctuations in the numbers of lynx and snowshoe hare in Arctic Canada are a good example: trapping returns of the Hudson Bay Trading Company show that cyclical fluctuations in populations of these predators (the lynx) and their prey (the hare) have occurred at approximately 10-year intervals over the past century, and that the predator cycle lags a year or two behind that of its prey. There is evidence, however, that the number of lynx has been too low to cause a rapid decline in the hare population every 10 years. What appears obvious is that some unknown factors, perhaps related to the food supply, are responsible for the hare cycles, and that the lynx population merely follows these sharp rises and falls rather than driving them along. The general absence of regular cycles in predator and prey populations under natural conditions can be accounted for in several ways. For instance, they could be masked by the many factors that affect most populations, as discussed in the previous chapter; or the predators may be *polyphagous* (that is, feeding on more than one kind of prey), in which case the interactions become very complex and difficult to analyze.

It is just such complexities that intrigue the ecologist. He wants to study not only how single populations expand and contract but how they intermingle with other populations. A single predator species and its prey are only part of a much larger system, where both prey and predator interact with many other plant and animal species. Some of the finest work that ecologists have done is concerned with the relationships among plant and animal populations in communities.

Simple System of Oranges and Rubber Balls

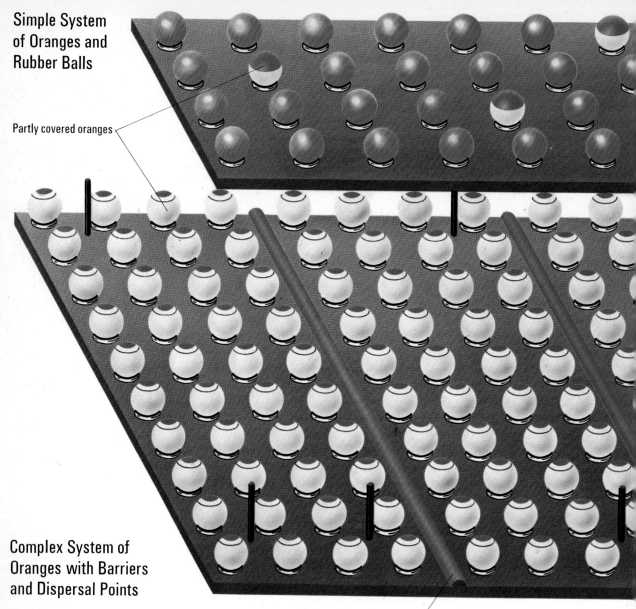

Partly covered oranges

Complex System of Oranges with Barriers and Dispersal Points

Paper bridges linking trays

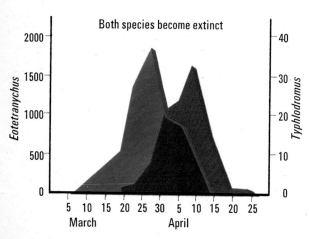

C. B. Huffaker studied predation under laboratory conditions using two species of mites as predator and prey. The populations in his simple system of oranges and rubber balls were unstable. Both increased rapidly at first but the prey were soon exterminated, whereupon the predators starved (graph left). In the much more complicated system of oranges with barriers and mite dispersal points, the predators were unable to eliminate all the prey at any one time and hence both populations persisted, showing regular oscillations (graph right).

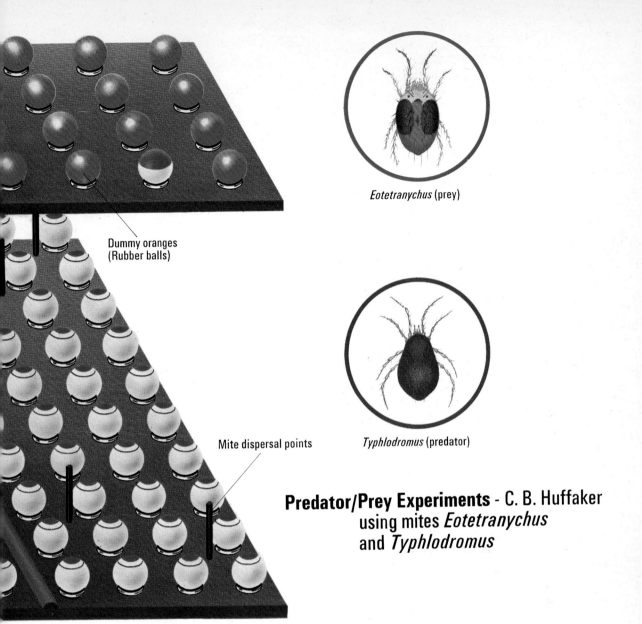

Dummy oranges
(Rubber balls)

Mite dispersal points

Eotetranychus (prey)

Typhlodromus (predator)

Predator/Prey Experiments - C. B. Huffaker
using mites *Eotetranychus*
and *Typhlodromus*

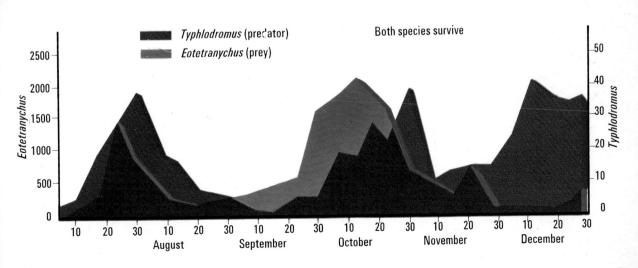

Typhlodromus (predator)
Eotetranychus (prey)

Both species survive

Food Webs
and Communities

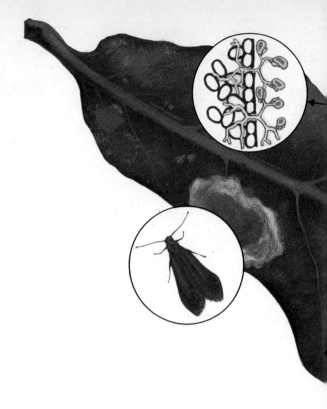

Ecologists speak of an assemblage of animals and plants within a given area, as well as their interactions with one another, as an "ecological community." The size of such communities can vary enormously. On a single leaf, for example, we might find plants such as fungi and various other microorganisms, herbivores such as aphids, book lice, and caterpillars, and predators such as insect parasitoids, bugs, and beetles. The community on an entire tree is much more complex than this, of course, but it is still relatively simple compared with that of a whole forest. A scientist may study the community of a leaf, a tree, or a whole forest, depending upon the scale of the investigation.

The concept of the "community" has long been recognized. As far back as 1877, the German biologist Karl Möbius wrote: "Every oyster bed is . . . a community of living beings, a collection of species and a massing of individuals, which find here everything necessary for their growth and continuance, such as . . . sufficient food . . . and a temperature favorable to their development." Obviously, Möbius realized that a community is more than a mere collection of species and that the organisms in any habitat interact with one another so as to form a more-or-less self-contained unit. Thus, the study of communities is one of the oldest branches of ecology, and its development has depended upon the patient work of many people. In this chapter we shall look at some of the studies that have been stepping-stones toward a broader understanding of community structure and function.

Our starting point must be an understanding of food webs, for these webs (which tell us "what eats what") are the fundamental functional units of all communities. Their complexity depends, of course, upon the number of species present. Within a termite mound, for example, there is essentially a two-species food web: the termites and their fungus. Worker termites bring plant material back to the nest, and upon this they grow a highly specialized fungus. This fungus is then grazed by the termites; their feces are either returned to the fungus garden or incorporated into the nest structure.

Such simple food webs are extremely rare, however. Even the smallest rock pool on the seashore is likely to contain dozens of interacting species. There may be frondose, filamentous, encrusting, and coralline seaweeds, upon which limpets, periwinkles, and various shrimps feed. These in turn serve as prey for dog whelks and other predators. Another, almost independent food web within the same pool, based on plankton, may consist of such plankton-eaters as acorn barnacles, anemones, and mussels. A link between the two webs is likely to be a predator such as the dog whelk, which eats barnacles and mussels as well as limpets and periwinkles.

When we examine very large communities, we quickly discover that the complete food web is too complex to be unraveled, and that the most profitable course is to focus on those links in the web that relate directly to only a single species. Perhaps the most noted example of a

A community of organisms is a group of populations interacting in the same place, whether it be on a small scale such as a single leaf, or large enough to include an entire forest. The oak leaf community shown here illustrates that many interactions can occur even in this small area. Feeding relationships are shown by the arrows, and the dotted circles indicate the species, such as the chalcid wasp, that are found on the lower leaf surface.

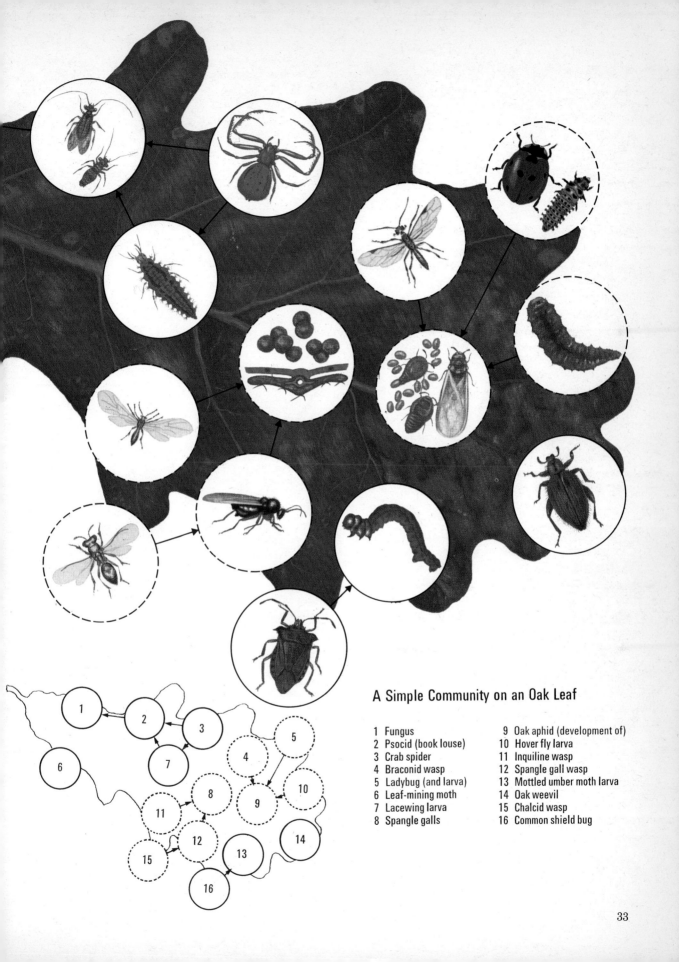

A Simple Community on an Oak Leaf

1 Fungus
2 Psocid (book louse)
3 Crab spider
4 Braconid wasp
5 Ladybug (and larva)
6 Leaf-mining moth
7 Lacewing larva
8 Spangle galls
9 Oak aphid (development of)
10 Hover fly larva
11 Inquiline wasp
12 Spangle gall wasp
13 Mottled umber moth larva
14 Oak weevil
15 Chalcid wasp
16 Common shield bug

detailed study of just one species is a project carried out in the early 1920s by a young naturalist, Alister Hardy (who was later to become Professor Sir Alister Hardy at Oxford University), working at the Lowestoft laboratory of the British Ministry of Agriculture and Fisheries. Hardy's investigation of the food of the North Sea herring throughout its life cycle culminated in the publication in 1924 of one of the most detailed food webs ever to have been drawn up as a result of observing a complex natural community. Before Hardy's study, it was known that herring feed exclusively on the plankton, of which they themselves are members during the early stages of their life cycle; but for the full details of precisely what eats what—and when—he deserves all the credit.

The essential characteristic of marine plankton is that it is composed of an astonishing variety of tiny animals and plants drifting freely and carried along by water currents; even though many of them can and do swim, they are not powerful enough to move against the general drift. An opposing term, the "nekton" (from the Greek *nektos*, meaning "swimming"), applies to all the marine animals that can swim strongly and are therefore independent of currents. A young herring remains in the plankton for several months before becoming a member of the nekton. Clearly, then, an essential requirement for Hardy's study was a dependable technique for taking a proper sample of the plankton, to assess the different kinds of animals and plants present and their relative abundance. Furthermore, the gut contents of some of the trapped animals could be analyzed to reveal the nature of a previous meal. It was possible to establish other feeding relationships by direct observations of living animals in aquariums.

The modern plankton net has been in use since 1828, and is still widely used for routine sampling. It is cone-shaped, with a glass jar attached to the narrow end, and is usually towed behind a slow-moving boat. What you catch in it depends on the size of its mesh. If the net has a very fine mesh to catch the very smallest members of the plankton, it must be towed at an extremely slow speed in order to prevent the net being burst. Such a net cannot catch the larger animals, because they are able to swim out of its path. These larger members of the plankton must be caught with a coarser-meshed net towed at a higher speed—and any such net, of

The variety of organisms caught in a plankton haul depends upon two major factors: the depth at which the net is fishing, and the size of the mesh. These two pictures show a simple plankton net in use at different depths. Left: a diver pulls a net through the water close to the sea floor outside a hydrolaboratory. Above: surface plankton are collected in a net towed by a boat.

course, misses out the smallest creatures. Whatever its mesh size, the net must be weighted to keep it from rising to the surface. When it is drawn into the boat, the plankton on the sides of the net are washed down into the jar, which is then removed from the net.

As well as such sampling in restricted areas, Hardy also needed records of the distribution of plankton throughout the North Sea. To tackle this with conventional plankton nets would have been a daunting task. And so he developed a more sophisticated plankton-collecting device, which allows a continuous record to be made of the plankton catch over hundreds of miles. This continuous plankton recorder looks rather like a wingless airplane, with a propeller at the tail end. The recorder is towed behind a boat at a depth of 33 feet—a suitable depth for picking up a representative cross section of the plankton. Water enters through an opening in the nose and any plankton in the water is filtered onto a continuously moving roll of gauze, which is immediately covered by a second strip making a "plankton sandwich." This is then wound onto a drum in a tank of formalin and so the plankton is preserved. The whole mechanism is driven by the propeller, which revolves as water flows over the blades. Thus a given length of gauze with its preserved specimens held in place by the "sandwich" always corresponds to a fixed distance traveled through an identifiable area of the sea.

This sampling device has proved so valuable that scientists today use it in pretty much its original form. The data it has provided through the years since 1924 have told us a great deal about how the abundance of plankton species has varied, and also about the rather patchy distribution of plankton over the Atlantic and the North Sea.

Although, at first glance, plankton collections look fairly uniform, they in fact contain a remarkable diversity of organisms. Among the plants are diatoms of all shapes and a variety of green flagellates (unicellular organisms with a long whiplike flagellum that propels them through the water). Among the animals are vast numbers of other unicellular organisms, small jellyfish and allied species, worms, crustaceans, and the larval stages of a great many different marine animals, including of course the herring. The food web that Hardy finally produced as the result of his studies involves about 20 species of animal upon which the herring feeds. These in turn are either predators that feed upon one another or herbivores that feed on a range of diatoms and flagellates.

At spawning time the herring collect together in enormous schools up to 10 miles long. Each female lays about 10,000 eggs, which sink to the bottom and stick to stones and other solid objects. The emerging transparent fry are about a quarter of an inch long and for the first few days they are nourished by the remains of the yolk within their yolk sacs. They then rise to the surface and become part of the plankton, feeding as herbivores upon tiny planktonic plants. At this stage, great numbers of them are eaten by predators, notably small jellyfish, arrowworms, the sea-gooseberry (a relative of the jellyfish), and *Tomopterus*, a swimming worm. As the surviving herring grow, they in turn become predators feeding on planktonic crustaceans, particularly two very common ones, *Calanus* and *Pseudocalanus*. When they are nearly two inches long, the herring undergo a rapid metamorphosis in which they acquire a true fishlike shape and a covering of scales, although they are still far from being mature. At first they

congregate in schools in estuaries, often along with a closely related species, the sprat (at this stage, the two species are collectively named "whitebait"). After about six months they disperse until they become fully grown, at which time they finally join up with huge schools of adult fish. The complete development of the herring from egg to mature adult takes three to four years. The prey of the adult is quite varied, and even includes the arrowworms that are a major predator of the herring's larval stages.

To concentrate on what the herring eats is to see only one part of the food web, for the whole web must also include the major species that eat the herring. Hardy found that predation of the herring begins not in its larval stages but even earlier: before the eggs can hatch, many of them are eaten by such predators as haddock. Indeed,

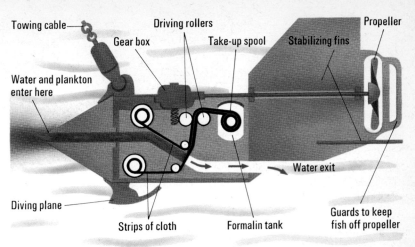

The continuous plankton recorder (far left and above) developed by Sir Alister Hardy provides a continuous record of the plankton caught over long distances. The recorder is towed behind a boat and the plankton (left) are collected between two strips of gauze, wound onto a drum, and preserved in a tank of formalin. When analyzed, these collections show where the young planktonic herring (shown below in different stages of development) are most abundant and also the distribution of the planktonic crustaceans on which the adult herring feed.

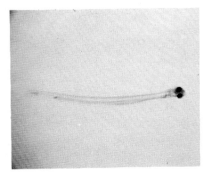

fishermen often interpret the presence of haddock gorged with fish eggs in a given stretch of water as indicating nearby herring breeding-grounds. As for adult herring, they are preyed upon by killer whales and porpoises, seagulls and gannets, codfish, man, and many other species. Cod, Hardy discovered, are the most important natural predators of the adult herring; in fact, 12 herring were found in the stomach of a single large cod. But the greatest impact on herring stocks has recently been overfishing by human beings. It is a feature of most food webs that they are linked together by a "top predator"—and man, alas, has become top predator in too many marine food webs.

An example of a terrestrial food web that has been investigated in some detail involves three plant species in abandoned farmland in the state of Georgia in America. We owe this to Professor Eugene Odum, of the University of Georgia, who made a study in 1962 of the food chains based on a sorrel, a sorghum, and a yellow daisylike plant called *Heterotheca*. What was particularly exciting about Odum's study was his very modern technique for determining not only what eats what, but also how much. He applied a radioactive tracer isotope to the plants by either spraying or injection, thereafter following the movement of the tracer up the food web. Thus he was able to establish the herbivores that fed on the three plants, such as grasshoppers, crickets, and ants, from the traces of isotope found within them. Having done this, Odum went on to look for further traces of isotope among likely predators of these insects. The predators proved to be predominantly wolf spiders.

As we have said, a great advantage of Odum's method is that it reveals not only the various members of a food chain but how much each has eaten, because the amount of isotope found in each species can be measured. Hardy did not have such a tool at his command when he made his herring study.

Despite the enormous variability among food webs, most of them share a basic structure. This fact was first recognized in 1927 by an Englishman, Charles Elton (later director of the Bureau of Animal Populations at Oxford). Elton's classic example was the simple community of any small pond, in which there are certain to be millions of protozoans, hundreds of thousands of water fleas, hundreds of beetle larvae, dozens of fish, and perhaps one heron. In almost all communities, therefore, each step up the food chain brings a decrease in abundance, which can be thought of as progressive steps up a pyramid—hence the *Eltonian pyramid of numbers*. In working out an Eltonian pyramid for a large community, we group together all individuals, irrespective of species, that occupy the same level in the food chain. Thus at the broad base of any large-scale pyramid stand all the green plants in the food web; above the base come all the herbivores, which feed upon the plants; and above them there may be several tiers of predators, culminating in such top predators as eagles, lions, and sharks.

Obviously, the top predators must be relatively scarce because there are decreasing numbers of prey at each successive stage of the pyramid. In between, however, there are sometimes exceptions to the rule of diminishing abundance. An example of a partly inverted pyramid can be found in a coniferous forest, where there are few trees in comparison with the vast numbers of caterpillars and other insects that feed on the needles. Above that level, though, the pyramid takes its usual shape.

So from the painstaking work of such pioneers as Hardy and Elton we have learned much about the functional units of natural communities— the food webs—and how they are composed. But a knowledge of the structure of food webs is only a beginning. For a fuller understanding of natural communities, we need to know how the energy in food is transmitted from step to step through the food web. Eugene Odum, as we have seen, used a method to discover quickly how much of a certain food is eaten by a given consumer. The

first really clear general exposition of energy transmission within the food web, however, was due to another young American ecologist, Raymond Lindeman, who died in 1942 at the age of 27. As a result of his detailed study of a small boggy lake in Minnesota, Lindeman was able shortly before he died to lay down theoretical foundations for the future study of feeding relationships in all communities.

Cedar Bog Lake is a typical small lake, with the usual populations of planktonic organisms, pondweeds, water fleas, and other small invertebrates and fish. Lindeman used Cedar Bog Lake and its inhabitants to show how any lake community can be stratified into a number of *trophic* (feeding) levels, which are the same as the different steps in the Eltonian pyramid of numbers. Plants, for instance, constitute the first trophic level, herbivores the second level, and predators the third and higher levels. In describ-

Radioactive isotopes have been valuable in studying food webs. In this experiment, the isotope was applied to the plants and its appearance in the herbivores and later in their predators was recorded. The relative concentrations of isotope in the various species of the same trophic level are shown by the degree of blue shading behind each circle and the major links in the web are indicated by white arrows. The technique demonstrates the relative importance of each link in the food web.

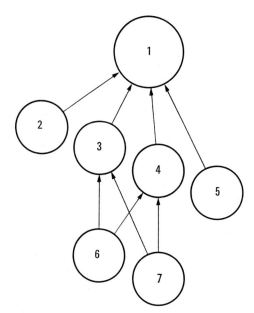

1 Plant *(Heterotheca)*
2 Snail
3 Grasshopper
4 Cricket
5 Ant
6 Wolf spider
7 Predatory beetle

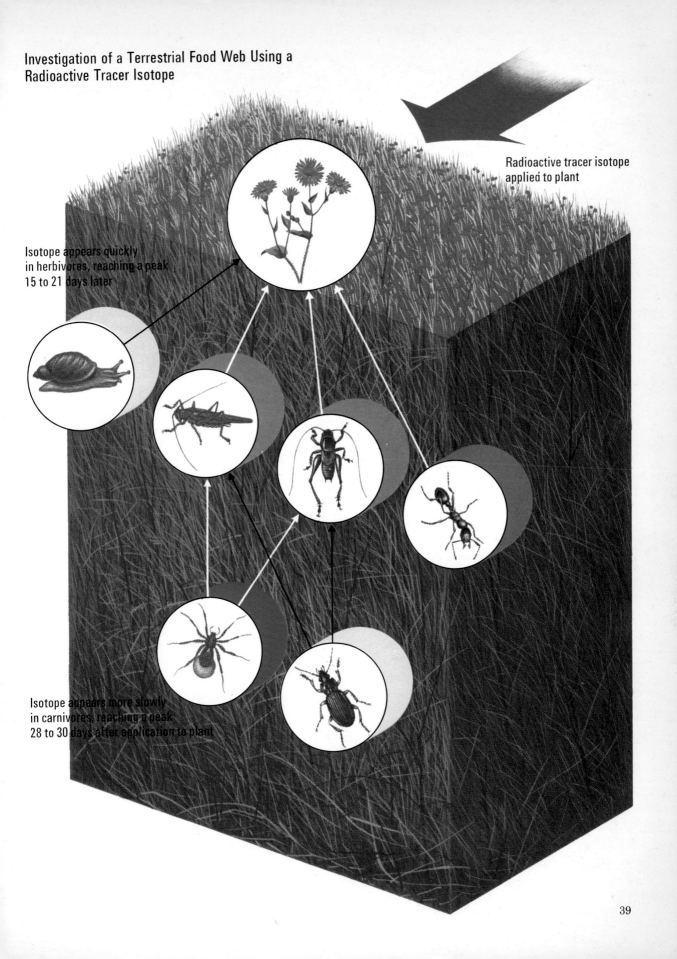

Investigation of a Terrestrial Food Web Using a Radioactive Tracer Isotope

Radioactive tracer isotope applied to plant

Isotope appears quickly in herbivores, reaching a peak 15 to 21 days later

Isotope appears more slowly in carnivores, reaching a peak 28 to 30 days after application to plant

ing the herring food web, Hardy had been concerned purely with whether or not species X feeds on species Y. Lindeman went much further; he was concerned with how much of Y (in calories or their equivalent) is eaten by X, and of this amount eaten, how much is excreted, respired, or utilized for growth and reproduction. Any amount that does contribute to X's growth and reproduction is of course then available as food for the predators of X. Stated very simply, Lindeman's objective was to evaluate exactly how much of the energy of Y at the foot of the pyramid is available to X, how much to the predators of X, and so on up to the top.

Let us consider in more detail the passage of energy from trophic level to trophic level in a typical lake. The ultimate source of energy for all life is the sun. Green plants absorb only a minute fraction (about 2 per cent) of the solar energy that reaches the earth, but this is enough to support the whole diversity of life on this planet. About half the solar energy absorbed by plants is respired—that is, used for self-maintenance—leaving only about 1 per cent for plant growth and reproduction. This amount, then, constitutes the community's *primary production*, representing the total amount of yearly growth in the lake's phytoplankton and pondweeds. It therefore provides the essential energy source for all the herbivores of the second trophic level. (Not all primary production is eaten, of course, because a sizable fraction dies and goes to a separate food web: the decomposers.)

The herbivores are either members of the zooplankton or larger grazing animals such as pond snails and herbivorous fish (carp, for instance). Just as not all the sunlight absorbed by plants is utilized for their growth and reproduction, so Lindeman found that most of the consumed plant material is merely ingested and then removed as feces. The food that is digested and absorbed may be either respired and used for body maintenance or allocated to body growth and reproduction (i.e. *secondary production*). Clearly, this is comparable to the situation discussed for the plants. Furthermore, it is a pattern that applies at all trophic levels.

To get an idea of the methods that Lindeman used to find out how much food is eaten at each trophic level, and how much of this is either burned off as respiration or lost as feces, let us look at another example: the winter moth

population discussed in Chapter 2. Let us suppose that we want to measure how much of the energy in the oak leaves on which the caterpillars feed is eventually incorporated into the winter moth population. There are several important stages in such a study.

First we burn a sample of oak leaves in a special instrument called a *bomb calorimeter*. This measures the energy content of the leaves, which can then be extended to the whole tree by estimating its total number of leaves. Secondly, we count the number of leaves eaten and estimate the number of caterpillars per tree by the method described in Chapter 2. This enables us to work out the approximate intake of food energy by the caterpillar population. Thirdly, we calculate the amount of energy lost as feces. We do this by collecting feces from a known number of caterpillars and measuring its energy content using a bomb calorimeter. Next, the energy lost by respiration of the caterpillar population is determined using a respirometer, which measures the oxygen used by the caterpillars over a known period of time. Each unit of oxygen consumed indicates a fixed amount of energy released. Finally, we determine the amount of growth and reproduction of the winter moths by weighing caterpillars at intervals and also weighing any eggs produced by adult females. These figures are then converted to calories by bomb calorimetry.

Our final results tell us just how much of the original energy from the primary producer (the oak tree) has become available to the predators of the winter moth. And by a similar set of measurements we can study the energy transfer at all subsequent trophic levels.

Lindeman found that, for the Cedar Bog Lake community, only about one thousandth of the solar energy that reached the lake in a year appeared as primary production by all the green plants, and that the herbivores incorporated only one eighth of this as growth and reproduction. At the next level, one quarter of the herbivore production was incorporated into predator tissue. The amount of food energy that reached the top of the food web was therefore a minute fraction (less than one 33,000th) of the amount initially available.

Several recent studies of relatively simple natural communities have confirmed the picture drawn up by Lindeman. One notable example was the study of Root Spring, a woodland spring in

Massachusetts, carried out in 1957 by Professor John Teal. The Root Spring community was an exceptionally simple one. The spring itself, no more than six feet in diameter and four to eight inches deep, contained few plants and only 40 animal species, of which only 12 species were really important for the overall picture of energy transfer. The main source of energy for the food web was dead plant material, such as leaves and twigs from surrounding trees and, to a lesser extent, the dead remains of duckweed and algae, which were the principal plants present. Neither of these plants was grazed by herbivores, so that the herbivore trophic level in Lindeman's community is represented only by detritus-feeding animals that aid in decomposition: worms, freshwater slaters and shrimps, mussels, snails, midge larvae, and caddis-fly larvae. The major predators were further midge larvae and flatworms.

Teal's first task was to estimate the amount of energy entering the system as dead plant material. To do this he placed wooden traps beside the spring, to catch and hold dead leaves and twigs falling down from the surrounding trees. From this evidence he was able to judge that about three quarters of the system's energy came from the trees, with the remaining quarter coming from dead algae and duckweed. By taking monthly samples of the animals in the spring, counting and weighing them, and then determining their energy content, he could make dependable estimates of the growth and reproduction of each species. He also determined experimentally the amount of food energy lost by respiration in each population. From all such detailed observations, he calculated that one ninth of the detritus was incorporated as growth and reproduction by the detritivores (as compared with one eighth by the herbivores in Lindeman's work), and that one fifth of the detritivore production appeared as predator tissue (as against one quarter in Lindeman's work).

The close correspondence between Teal's and Lindeman's figures certainly supports the impression of an underlying general pattern in the way food energy is transferred from one trophic level to the next. But it is very hard to confirm any such hypothesis from studies of natural communities, because natural communities are generally too complex and variable to permit accurate measurements. Thus, as with so many ecological questions, it becomes profitable to investigate them in the laboratory, where carefully controlled experiments can be carried out on artificially simplified systems.

The most important laboratory tests of Lindeman's figures that have been made are those of Professor Lawrence Slobodkin of the University of Michigan, who devised an experimental system of three trophic levels, each with only a single species. Plants were represented by a tiny, unicellular alga, herbivores by a water flea, and the predator by Slobodkin himself (whose predation consisted in removing the fleas from the community). The water fleas were placed in glass beakers and supplied with algal food at four different controlled levels of abundance. Slobodkin managed this by growing the algae uniformly on plates of nutritive agar jelly and then carefully varying the amounts of agar added to each beaker. He also varied the predation level by changing the proportion of fleas that he removed. In this way he was varying both the input of energy from the plants and the output of energy due to predation.

Slobodkin soon found that his figures for the amount of material converted into water flea growth and reproduction varied only with the intensity of predation, *not* with the level of abundance of plant food. Furthermore, the highest figure was almost one eighth, much the same as the results of Lindeman and Teal, which supports the idea that natural communities have evolved to make the most efficient use of food energy. When Slobodkin repeated his experiments with another simple system, using brine shrimps as prey for hydras, with himself as top predator, he again obtained similar results. On the strength of this he suggested that an efficiency rate of approximately 10 per cent in the transmission of energy from one trophic level to the next is universal in natural communities.

In fact, valuable as Slobodkin's work was, this sweeping generalization is not entirely valid, for the figure applies generally only to aquatic systems. The freshwater communities upon which Lindeman and Teal worked are significantly different from many terrestrial communities in their pattern of energy transfer. The major difference between aquatic and terrestrial systems is that warm-blooded animals—mammals and birds—are much more active as grazers and predators on land than in water. This single difference is important to the whole food web simply because of the energy costs of maintaining

Sunlight

Dead Leaves
and Twigs

Respiration

Detritus and Algae

Respiration

Herbivores

Respiration

Carnivores

a high body temperature. Cold-blooded animals burn only about half the energy in their digested food as respiration, whereas warm-blooded animals use up as much as 98 per cent in this way. Thus, the figure of approximately 10 per cent for the energy transfer between trophic levels cannot apply where mammals are an important part of the food web. In these cases an efficiency rate of one per cent would probably be nearer the mark, which is why terrestrial communities cannot generally support as many trophic levels as do their aquatic counterparts.

The studies of ecologists are not merely of scientific interest but have a direct relevance to practical problems. When we apply their findings to an analysis of man's production of food, we become aware of a striking fact: compared with the feeding habits of other animals, our ways of producing food energy are extremely inefficient. It is a revealing exercise to examine six steps in human food production. As pointed out by Dr. John Lawton of the University of York, a certain amount of energy is used up at each of these steps, and we must balance these losses against the

Teal's Study of the Energy Flow in a Woodland Spring

1 Algae
2 Coarse detritus
3 Fine detritus
4 Hog louse
5 Caddis-fly larva
6 Pea shell
7 Freshwater shrimp
8 Bladder snail
9 Midge larva
10 Worm
11 Caddis-fly larva
12 Midge larva (predatory)
13 Flatworm

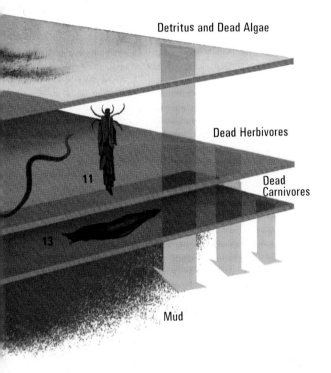

Detritus and Dead Algae

Dead Herbivores

Dead Carnivores

Mud

This diagram illustrates the results of a study of a small woodland spring in Massachusetts. The arrows represent the flow of energy between the various parts of the system, with the relative size of the energy flow shown by the width of the arrows (not to scale). Energy enters the system as sunlight and—more importantly in this example—as dead material from the surrounding trees and other plants. The layers represent successive trophic levels, and the energy that enters each one is balanced by losses due to respiration, dead material going to the bottom mud, and consumption by the next trophic level.

energy that is gained from food when it is eaten.

Step 1 consists of all the energy expended in growing crops and rearing livestock. It includes the running of farm machinery for plowing and sowing and the manufacture and application of fertilizers and pesticides. Of these, fertilizers are by far the most energy-expensive. The only noteworthy animals, other than man, that use fertilizers are the ants and termites that cultivate "fungus gardens." But because they have no machinery and recycle their own fertilizer in the form of feces, their efficiency at this stage is much greater than ours. Similarly, the few species of insects that rear livestock do so far more economically than we do. Best-known are the ants and their aphid "cattle" from which they get a sugary liquid called "honeydew." Some ants not only tend aphid colonies and ward off predators, but also move their "herds" from one plant to another, and even harbor them in their nests in wintertime.

In *Step 2* John Lawton considered the energy expended in searching for food supplies. This is a negligible operation in modern agriculture, although it remains important for such communities as the Eskimos and African nomads. On the other hand, it is a major step for most non-human animals.

Step 3 consists of the energy used in harvesting plants and killing animals. Harvesting involves enormous costs in modern farming; think of the vast quantities of fuel that are needed for running combines. This is also an important step for other animals, especially for those that must chase, catch, and kill active prey. Their net gain of energy, therefore, is the balance between what they get from eating a prey and what they use for its capture.

Step 4 includes the energy expended in food transport and storage. This, too, is vital in modern agriculture, especially because so much food today is exported around the world and stored for long periods. Although such animals as pikas, squirrels, and bees store food, the process takes only a very minor part of their total energy expenditure.

In *Step 5,* John Lawton considered the energy used in processing and cooking food. In making bread, for instance, the grain is first milled, then made into dough, and baked. We use fossil-fuel energy at each stage, and especially in the heating of ovens. The equivalent in other animals might be the cow's chewing of the cud, the squirrel's cracking of nuts, or the thrush's cracking of snail shells. But all these natural processes involve only insignificant energy costs.

Finally, in *Step 6* there is the energy used in the chewing and swallowing of food—a negligible factor in all cases, with the possible exception of such extremes as the effort of a boa constrictor swallowing a very big prey.

At steps, 1, 3, 4, and 5, then, modern agriculture is energy-demanding to an extent unparalleled in the animal kingdom. It is even probable that some types of agriculture are operating at a net energy *loss*. In other words, more fossil-fuel energy is sometimes expended in the job of producing our food than we gain in energy when we eat it. Only in steps 2 and 3 do most non-human animals "waste" significant amounts of energy. Surely we could begin to curb our own extravagances if we gave more thought to what the ecologists have learned about energy transmission within food webs.

One aspect of modern farming of particular interest to ecologists—and of incalculable importance to the welfare of all of us—is the use of persistent pesticides that tend to be concentrated as they move up from one trophic level to another. To see how DDT, for example, can affect a large-scale ecological community, let us look at the findings of an important study carried out in the 1960s by Dr. George Woodwell of the Brookhaven National Laboratory, and Dr. Charles Wurster and Dr. Peter Isaacson of the State University of New York. They set out to determine the level of DDT in different parts of a salt-marsh community on Long Island, New York, that had been sprayed against mosquitoes for the previous 20 years. The results were spectacular. Up to 32 pounds of DDT residue were found in each acre of the upper layer of mud—an accumulation that inevitably occurs with DDT, because it breaks down very slowly in the environment. Much more disturbing, though, was the way the DDT became increasingly concentrated toward the top of the food chain. Green plants such as the common algae of the marsh contained 0.08 parts per million (a relatively low concentration), and the total plankton had an even lower concentration of 0.04 parts per million. The animals that fed on the plankton—mosquito larvae, snails, clams, shrimps, eels, and other fish—contained between 0.16 and 1.28 parts per million of DDT, a concentration of up to 32 times as much as in the lower trophic

Modern agriculture uses vast amounts of fossil-fuel energy, but animals must rely on their own muscle power for finding and processing food. The essential difference is that man, by substituting machines (above) for muscle, has freed much of the population from the task of food gathering. Most animals must spend a great deal of time and energy in gathering and processing food. A thrush, for example, collects large numbers of snails and takes them to a convenient stone, or anvil (left), where it breaks open each shell and then eats the snail.

level. The next level, consisting mainly of such predatory birds as gulls, cormorants, herons, ospreys, and terns, had concentrations within the body tissues of from 3 to 75 parts per million— as much as 1000 times the amount found in the primary producers of the salt-marsh food web.

The mechanism of concentration is simple. Very little DDT is broken down at each stage in the food web; thus nearly the same quantity that is divided among many primary producers at the base of the Eltonian pyramid is divided among a smaller number of consumers on the next trophic level, and so on up to the apex, where the entire amount is concentrated in the tissues of a very limited number of top predators.

45

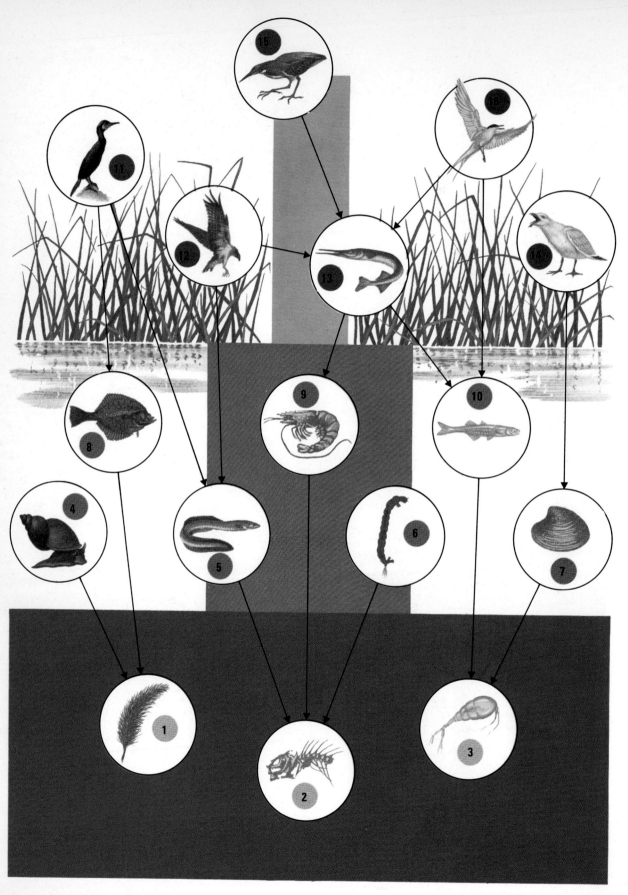

The Effects of DDT on a Salt-Marsh Community

1 *Cladophora*	9 Shrimp
2 Organic debris	10 Silverside
3 Plankton	11 Cormorant
4 Snail	12 Osprey
5 Eel	13 Billfish
6 Mosquito larva	14 Gull
7 Clam	15 Heron
8 Fluke	16 Tern

This diagram shows the relative biomass of three trophic levels in a salt marsh on Long Island Sound together with some of the major links in the food web of the marsh. DDT applied to the lowest trophic level in the pyramid is passed with little loss to successive trophic levels and hence is concentrated within progressively smaller biomasses. The degree of red coloring of the dots within the circles indicates the relative concentration of DDT accumulated in each species. Below: a western grebe carrying its young. This species was considerably affected by applications of another pesticide, DDD, to Clear Lake, California.

The highly toxic nature of DDT has led to some unfortunate side effects, even when the concentration is less than a lethal dose. Experiments in which chickens, pigeons, quail, and other birds have been fed various amounts of DDT and other pesticides have had such disquieting results as a reduction in the number of eggs laid, failure of eggs to hatch, and a reduced survival rate for chicks. The levels at which these effects occur in the case of DDT are in the region of 30 parts per million—well within the range of concentration found in the Long Island salt-marsh birds.

Perhaps the best-known instance in recent history of a major ecological upset resulting from the build-up of an insecticide in a food web is that of Clear Lake, 100 miles or so north of San Francisco, California. DDD, which is a compound closely related to DDT, was first used there in 1949, when the midge population became so numerous as to be a serious nuisance to fishermen and holidaymakers. The authorities chose to use DDD because they thought it would be less harmful to fish than DDT, and they expected it to sink to the bottom of the lake, where it would kill the aquatic larvae of the midge. The immediate effects delighted everyone. The midges were almost completely eliminated and there was virtually no apparent damage to either the fish or such birds as the western grebe. In 1952, however, when hordes of midges reappeared and further applications of DDD were made, serious side effects were quickly apparent. Widespread deaths of the western grebe were observed, and there was certainly a great reduction in their breeding success—a reduction that had probably been going on unnoticed for some time.

Scientists soon discovered that the concentration of DDD in the grebe was even more spectacular than that of DDT in the salt-marsh birds. Only 0.015 parts per million of the compound had been introduced into the water of Clear Lake. Yet the plankton contained 5 parts per million; plankton-feeding fish contained about 10 parts per million; predatory fish contained an even higher level; and the grebes, at the very top of the food web, contained as much as 1600 parts per million within their body fat.

It is clear that the ecological study of food webs and communities has considerable importance for our own well-being. It enables us to manage our food resources more efficiently and to assess the potential hazards of pesticides and fertilizers.

Island Biogeography

Biogeography is the study of the distribution of animals and plants, and therefore represents an area of common interest to ecologists and geographers. One aspect of biogeographical studies that has had a special place in the development of biology is island life. Both Darwin and his contemporary A. R. Wallace were fascinated by the diversity of animal and plant forms that had evolved on islands—a fascination that in part led to their formulation of the theory of evolution by natural selection. Islands have continued to be of great interest not only to biologists but also to geologists and geophysicists. Although modern biologists still study plant and animal evolution on islands, they are perhaps more interested in such problems as the ways in which islands are colonized by animals and plants in the first place, and what determines the number of species that an island can support.

Conventionally, islands are divided into two categories: continental and oceanic. Continental islands have been separated from a continental landmass of which they once formed a part: typical examples are Great Britain and Japan. Oceanic islands, on the other hand, have never had a land connection with a continent, and either are of volcanic origin or were formed from the fringing reefs remaining after a volcanic island sank below the water's surface. Geologically, therefore, there is a clear distinction between these types of islands. The different methods of island formation have an important effect on the type of animals and plants found on them. A continental island will have started off with communities of plants and animals much like those of the mainland to which it was connected. Some of the species, however, will have been unable to survive under the new island conditions, leading to simpler communities with fewer species than the nearby continents. An oceanic island that started as molten rock would obviously have a very different history, as it gradually acquires all its species by means of air or sea transport or local evolution.

A series of experiments that highlight the problems of island colonization was undertaken in 1963 by Professor Basset Maguire of the

The Galápagos Islands are oceanic islands, which means that they have never had a land connection with a continent. Because of their isolation, they are characterized by a relatively large number of unique species. The marine iguanas lazing on the rocks in this picture are found only on the Galápagos Islands, as are the red land crabs that share their habitat on the rocky shore. Blue-footed boobies are among the many oceanic birds that have made islands their permanent nesting places.

University of Texas. He set up a number of "is-lands" (represented by small jars of sterile water) placed at various distances from a "mainland" (represented by a freshwater pond). The aim of the project was to record colonization of the jars by microscopic organisms from the pond. This would be an interesting study for an amateur ecologist to make. All you need do is set up a grid of jars at varying distances from a convenient pond, and then count the number of micro-organisms present in each jar at regular intervals of, say, a week. The likely result is that the number of species in each jar will increase rapidly at first, then more and more slowly until, after a few weeks, the number will remain more or less unchanged. Such a curve of species present in a given place over a period of time is called a *colonization curve*. As you might expect, one of Maguire's findings was that the number of kinds of organism finally present in the jars decreased with increasing distance of these "islands" from the "mainland." This can be accounted for by a lower rate of arrival, or immigration, of microorganisms at the more distant jars, and it throws light on why isolated islands contain fewer species than others: in general, living creatures have less chance of successfully dispersing over long distances than over short distances.

The way in which species reach islands—their dispersal mechanism—is of interest to both geographers and ecologists. Dispersal can be either active or passive. Active dispersal to islands is usually restricted to those groups that have strong flight, such as certain insects, birds, and bats. Even with such groups, however, dispersal is occasionally passive, as when strong winds blow the animals off course. Purely passive dispersal occurs in many groups of animals and plants, and can occur in a variety of ways—by air, by sea, or by transport on or in the bodies of other organisms. Just as the sea supports a vast floating population of small organisms—the *plankton*—there is also an aerial plankton made up of pollen grains, the spores of microorganisms, large numbers of small insects, spiders, and mites, and other living things. This aerial plankton is carried at the mercy of the winds at considerable heights all over the world, as was well shown in an experiment made in 1961 by Dr. Lindsey Gressit of the Bishop Museum, Honolulu.

Gressit fitted special traps to commercial airplanes crossing the Pacific. These devices were designed to catch the larger animals of the aerial plankton. More than six cubic miles of air passed through the traps, in which were caught 1075 insects along with some spiders and mites. Although the great majority of the creatures had perished from cold or desiccation, enough of them were alive to have provided a considerable number of colonizers wherever they might have landed. The distance involved in this dispersal can be great. For instance, when large numbers of aphids that attack spruce trees were found on the treeless Spitsbergen Islands in 1924, it was thought that they had drifted over 800 miles in the air across the Arctic Ocean.

Some living things can reach islands only by passive transport on the sea. Coconuts, for example, have evolved a fruit that can float for several weeks. The effectiveness of this means of dispersal is apparent from the wide coconut distribution on many remote islands of the Pacific and Indian oceans. But there are numerous ways of hitching rides to new places of habitation. Rafts composed of floating vegetation make good transport for a small beetle (*Microlymma*) that normally inhabits seaweed on the beaches of the eastern seaboard of North America. In this way it has been carried right across the Atlantic and has spread throughout the shores of northern Europe.

Finally, passive transport on the bodies of larger animals is sometimes the means by which organisms are dispersed. The mud on birds' feet, for example, may carry small seeds, the eggs of fish and other animals, adult microorganisms, and even insect larvae or small crustaceans. If a bird lands momentarily on a far-off island, one result of the visit may well be the addition of a new species to the island's community.

Some further light was shed on colonization curves, a few years ago, by two of America's foremost island ecologists, Professor Edward Wilson of Harvard University and Professor Daniel Simberloff of Florida State University. For their experiment they chose several tiny mangrove islands of the Florida Keys. These ranged in diameter from 12 to 19½ yards, and their distances from the nearest major source of immigration varied from 2 to 287 yards. Each of them was almost entirely covered by red mangrove shrubs, with occasional small bushes of black mangrove. The aim of Wilson and Simberloff's study was to record the number of insects, spiders, and other arthropods on each island, to remove them completely, and then to follow the subsequent colonization curves over a period of two years.

They successfully accomplished the first part of their experiment—a survey that revealed between 20 and 43 species of arthropod on each island. The bulk of these were insects, with a few spiders, scorpions, pseudoscorpions, centipedes, millipedes, and tree-dwelling sowbugs. These preliminary studies showed, too, that the islands nearest to the "mainland" tended to have the greatest number of species—a finding very much in accord with the results of Maguire's experiment with the jars and the pond. The first efforts to rid the islands of their arthropod populations, however, were unsuccessful. Wilson and Simberloff sprayed insecticides such as parathion and diazinon over each of their islands, but soon discovered that these had little, if any, effect on several species. In particular, some beetles, wasps, and ants within hollow twigs escaped the insecticide action.

Eventually they developed a more successful (though much more elaborate) fumigation technique, which involved the erection of scaffolding covered with a plastic-impregnated canvas tent around each island, and the release of methyl bromide gas into the tents. When the tents were removed after more than two hours, a thorough search for living arthropods revealed only one or two beetle larvae that had bored deep into the mangrove trunks and so sometimes escaped the fumigant. Such results were felt to be good enough to warrant their taking the next step in the experimental program: a sampling procedure that required an exhaustive search of each island once every 18 days. This sampling operation was carried out for about one year, and the search was then continued at less frequent intervals for the following year.

The results provided colonization curves for the arthropods of the same general form as those that Maguire had obtained for microscopic organisms. At first, new species accumulated rapidly on each island, but successful colonization became less frequent, until over the second year there was almost no further increase in species numbers. Of special interest was the finding that this *equilibrium* level of species on each island was very close to the initial level that had prevailed before the fumigation with methyl bromide. This provides good evidence of the speed with which an island community can attain its equilibrium complexity in terms of number of species.

Experiments such as that of Wilson and Simberloff have been very useful in developing a general theory of island colonization and species equilibrium levels. At the forefront of this theoretical work has been the late Professor Robert MacArthur of Princeton University, working with Professor Edward Wilson. It is their conclusion that the number of species present on an island depends on the balance between the rate of arrival of new species (immigration) on the one hand and the rate of loss or extinction from the island on the other. Moreover, the rate of immigration depends upon the distance over which the immigrants have had to disperse. Thus, for example, an island close to the mainland can be expected to have a higher immigration rate for both animals and plants than will an isolated island. The level of extinction, however, mainly depends on island size. A small island can support only relatively small populations compared with those of a larger island, and hence there is a greater risk of extinction. The equilibrium number of species on an island will occur when immigration and extinction are in balance. Thus, a small and isolated island, with its higher rate of extinction and lower rate of immigration, will support a smaller number of species in comparison with those of a large and less isolated island.

In the two basic experimental studies that we have described—Maguire's "island" jars and "mainland" pond and the Wilson and Simberloff mangrove islands in the Florida Keys—the sizes of the islands, and hence their extinction rates, were constant. Their distances from the nearest source of colonizers, and thus their immigration rates, were intentionally varied. In each case, the nearest islands, which had the highest immigration rates, were inhabited by more species at the end of the experimental period than the distant ones, with their lower rates of immigration. The experimental results are therefore in close agreement with the MacArthur-Wilson general theory of island colonization.

When we turn to natural examples rather than controlled studies, we find considerable support for these ideas. For example, the effect of an island's distance from the source of its colonizers is well illustrated by the relative abundance of weevil species on a chain running eastward into the Pacific from Australia and New Guinea. An island such as New Caledonia, which lies fairly close to the Australian coast, has 37 of these beetles, whereas in the Marquesas, lying much farther out in the Pacific, we find only seven. As a further example, we find support for the hypothesis that a large island can support more species than can a smaller one from the plant species on the different Galápagos islands. The largest of these, with an area of over 2000 square miles, contains 325 species of plant, whereas the smallest, of just one fifth of a square mile, has only seven species.

The most spectacular instance of island colonization can be found where volcanic activity has created new islands—Surtsey off Iceland, which was formed in 1963, is a good example—or has obliterated all life, as in the Krakatoa eruption of 1883. Krakatoa, Verlaten,

Colonization Curves for "Islands" of Various Sizes

The invasion of new habitats by plants and animals shows a common pattern irrespective of the kind of habitat or the species involved. During the early stages of colonization, new species are acquired rapidly, but the rate slows down as the habitat fills. This is clearly shown here by the experiments of Maguire and of Wilson and Simberloff. In the case of the colonization of Krakatoa by birds, an S-shaped curve developed because the major bird species could not invade the island until forests had developed.

The colonization of Krakatoa after the 1883 eruption

Wilson and Simberloff's Experiment:
colonization of mangrove islands
after fumigation

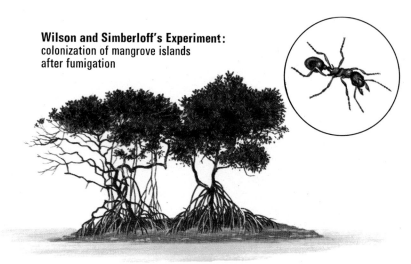

Maguire's Experiment:
invasion of jars of water
by microorganisms

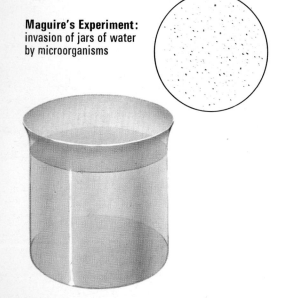

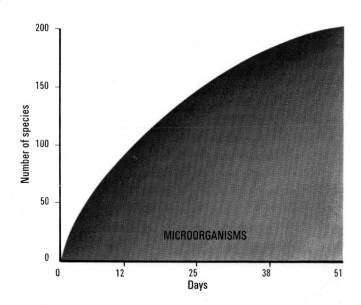

MICROORGANISMS

Number of species

Days

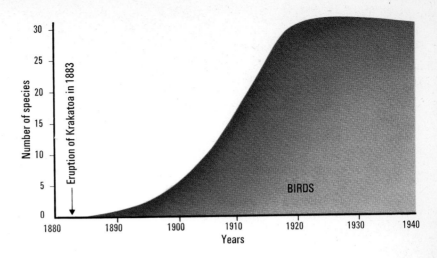

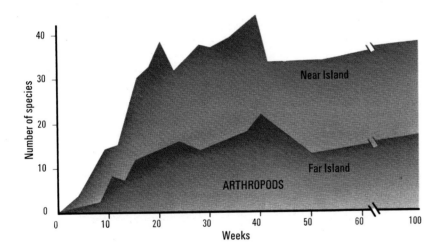

and Lang are three small islands lying midway between the continental islands of Sumatra and Java. In August 1883, Krakatoa erupted fiercely, with an explosion audible more than 2000 miles away, producing tidal waves (*tsunamis*) that were over 100 feet high and that traveled at a speed of over 50 mph. About two thirds—some eight square miles—of the island disappeared into the sea, and a layer of hot pumice and ash between 100 and 200 feet thick covered all the remaining land. We know little of the plant and animal life on Krakatoa before this time, but they certainly all perished in the eruption. During the following 50 years, three detailed surveys of life on the islands took place: in 1908, in 1919–21, and in 1932–4. The results of these

were summarized by the Dutch scientist, Dr. K. Dammerman in a publication in 1948.

In 1908, only 25 years after the disaster, an appreciable fauna had already begun to colonize Krakatoa. Among the 13 recently established species of birds were pigeons, kingfishers, and orioles; monitor lizards and geckos had also arrived, probably as passengers on floating debris from Sumatra or Java; and there were nearly 200 kinds of insects, mainly ants and flies, as well as some spiders, scorpions, sowbugs, and snails. The predominant vegetation was grass, but by the time of the 1919–21 survey a mixed forest had already started to replace the grassland. By 1919–21, too, over three times as many kinds of animals lived on the island,

MacArthur and Wilson's Theory
of Island Colonization

MacArthur and Wilson's theory states that the rate of immigration of new species to an island depends on the distance of the island from the mainland, whereas the rate of extinction depends on island size. The balance between these two will determine the final number of species, as shown in the brackets.

Mainland
(source of new species)

Near Island
(6 species)

Mid Island
(5 species)

Large Island
(7 species)

Effect of Distance from the Mainland

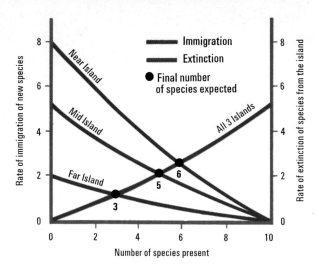

Effect of Island Size

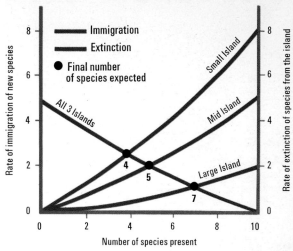

Small Island
(4 species)

Far Island
(3 species)

including brown rats (which had probably arrived by boat), two species of bats, new types of birds, lizards, and insects, and an abundance of earthworms. When Dammerman made the third survey in 1932–4, he and his fellow-workers found that most of the grassland had given way entirely to forest vegetation. Although some of the animal species previously recorded had disappeared, there were now 1100 species on the island, as compared with 268 species in 1908 and 573 species in 1919–21.

Clearly, the most rapid colonizers are those groups with strong flight, especially the birds. Yet the number of bird species on Krakatoa increased slowly at first—from none in 1883 to only 13 a quarter of a century later. By 1919–21 the number had jumped to 31, but only 30 species were recorded in the third survey. This gives a somewhat S-shaped colonization curve, rather different from the convex curve we might have anticipated. The rather slow bird colonization at first is due to the absence of a complex plant community of mixed forest. Only when this had evolved on the island could many of the bird species become established.

The example of Krakatoa serves to remind us that although such simple controlled experiments as those of Maguire can answer specific questions and help to develop theories, the real world is much more variable and complex. Ecology is best served when experimental studies and careful observation of natural systems at work are combined to produce general theories of the sort propounded by MacArthur and Wilson.

Classification of Habitats

The classification of animals and plants has been recognized as an integral part of biology for over 2000 years, since the time of Aristotle and his pupil Theophrastus. These two Greek philosophers, whom we consider to have been the founders of biological science, were great believers in organization and logical order, and, among their many other remarkable undertakings, they strove to provide an ordered framework in which to classify animals and plants. Interestingly enough from the standpoint of modern ecology, they based their system of classification on a description of different types of habitat, each of which contains characteristic kinds of animal and plant life. Aristotle divided all animals, for example, into water-dwellers and land-dwellers, and then subdivided each of these two major groups into many smaller categories. Thus, water animals could be either sea-dwellers, river-dwellers, lake-dwellers, or marsh-dwellers, and then could be further subdivided into those that are entirely aquatic and those that live and feed in water but breathe air and reproduce on land.

After its inspired start in ancient Greece, biology remained static for a very long time. There were almost no conceptual advances over the ideas of Aristotle until the late 17th century, when there was a new awakening in biology due to the works of such pioneers as the Dutch naturalist Anton van Leeuwenhoek, who invented the microscope, and Carl von Linné (Linnaeus), the 18th-century Swedish botanist who founded the modern science of classification, or taxonomy. In the 19th century, two noted English biologists, Charles Darwin and Alfred Russel Wallace, gave the classification of living things a new meaning as viewed in the light of their work on the evolution of species.

Today, taxonomy remains an unchallenged keystone of biology. Consider, for example, an ecologist studying a complex community such as a tropical rain forest. The success of his work will hinge upon his precise knowledge of the kinds of animal and plant that live in the community, and also upon his skill at distinguishing among the different types of habitat to be found there.

The immense variety of organisms means that the task of classifying them is a job for specialists. Indeed, when you realize that there are about 20,000 different sorts of alga, 250,000 mosses and liverworts, 250,000 higher plants, and a much greater variety of animals—including, for example, 10,000 sponges and nearly a million kinds of insects—it is not surprising that most specialists in taxonomy restrict themselves to particular groups. The classification of habitats, though not nearly so complex, is of particular importance to the ecologist as a shorthand way of identifying different environments and the types of plant and animal associated with them. The scale of this habitat description can vary enormously, of course—from a leaf to a major part of a continent, from a drop of water to a lake. On the larger scale, the habitats are usually distinguished from one another by such features as climate and indigenous plant life. Thus, on the one hand, there are the major biotic divisions, such as deserts, oceans, ice caps, grasslands, and so on; and on the other, there are much smaller-scale habitats such as a woodland, which may be subdivided into the canopy, the shrub layer, the field layer, and the ground layer.

A pioneer student of the large-scale classification of habitats was an American physician and biologist, Dr. C. Hart Merriam. In 1889, Merriam began a biological survey of an area in Arizona, and became particularly interested in the pronounced bands of distinctly different types of vegetation at successive heights on the 13,000-foot-high San Francisco Mountain. These bands seemed to him to repeat on a small scale the much larger zones of climate and vegetation encountered as we progress from the equator toward the high latitudes of the Arctic.

This interest led Merriam to develop a general vertical classification of habitats. It begins with a region of hot deserts, which he called the Lower Sonoran Zone after the Sonora Desert of

This scene illustrates the complex vertical structure of the vegetation found in a typical deciduous woodland. It shows at least three distinct layers of vegetation—the field layer, the low canopy, and the high canopy—each of which has particular animal and plant species associated with it.

Southwestern North America. This region is characterized in America by an abundance of cacti such as the saguaro or organ-pipe cactus, various prickly pears, the chola, which is a tree-like jointed cactus, a wealth of such yuccas as the Joshua tree and Spanish bayonets, and thorny shrubs such as mesquites and paloverdes. As we climb upward to about 4500 feet, we leave the Lower Sonoran and enter the Upper Sonoran, a region of dry-scrub woodland in which small trees and tough shrubs of many different kinds abound. Then, at 6500 feet, on the lower slopes of the mountain itself, rainfall increases markedly and we enter the first of the true forest zones. Here, in what Merriam called the Transition Zone, ponderosa pine predominates.

Beyond 8000 feet the vegetation changes once more, and we find ourselves in a cool, dark coniferous forest, which is so much like a typical Canadian forest, even to the extent of sharing many of the same trees—Douglas firs, lodgepole pines, and spruces—that Merriam named it the Canadian Zone. Similarly, above 10,000 feet lies a region of cold, wet forest reminiscent of the shores of Hudson Bay—hence the Hudsonian Zone, whose dominant plants are the Engleman spruce and bristlecone pine. (The bristlecone pine, incidentally, is a remarkable tree, with the distinction of being the longest-lived organism known on earth; some living specimens are known to be more than 4000 years old.) The upper limit of the Hudsonian Zone is the tree line at about 12,000 feet. Above this there stretches a community of dwarf shrubs and other low-growing vegetation of the kind normally found in the tundras of the true Arctic. Indeed, many of the species, such as certain saxifrages, gentians, and buttercups, are the same.

Merriam also studied the animals, particularly the mammals within each zone, to get some idea of their distribution. He found, as you might expect, that many species—most mammals, in fact—are primarily associated with particular vegetation zones. Their full distribution, however, often spreads over two or more adjacent habitats. The striped skunk, for example, is mainly an Upper Sonoran species, but can be found in lesser numbers throughout the Transition, and occasionally as high up as the lower parts of the Canadian Zone. In much the same way, there is some mixing of plant species between adjacent zones. Aspens, for example, which are most at home in the Transition Zone, also

thrive throughout much of the Canadian Zone.

The special significance of Merriam's work was that his classification of the San Francisco Mountain zones has also a much broader application. His conjecture that the vertical zonation on a mountain range mirrors continental vegetation zones from south to north has been borne out by later studies of the plant communities in different parts of North America. The similarity is so great that a useful rule of thumb is that an increase of 1000 feet in altitude is roughly equivalent to moving about 300 miles north in latitude.

Since Merriam's time, there has been a steady interest in the mapping and classification of the major plant communities of the world. On a global scale, this has culminated in the division of the land surface into a number of regions, or *biomes*, distinguished by their characteristic types of vegetation.

No exact number of recognized biomes has been agreed upon; different schools of ecological thought have different ideas. A basic list, though, would have to include about 10 distinct types. These are, moving roughly southward from the Arctic to the equator: tundra, northern coniferous forests (or *taiga*), temperate deciduous forests, temperate (or steppe) grassland, chaparral, desert, savanna grassland, tropical scrub forest, tropical deciduous forest, and tropical rain forest. Running south of the equator, of course, we find similar biomes in reverse order, with certain variations imposed by geographical differences in the southern latitudes. All terrestrial plant communities can be fitted into the pattern. Each biome tends to have a characteristic climate, which has shaped the dominant types of vegetation. The plant communities in turn provide special habitats for many specialized kinds of animals that feed on the plants and shelter in them.

Our knowledge of the major biomes is based on the work of a great many people. The early syntheses were largely made between 1909 and 1912 by such workers as the Danish plant ecologist Professor Eugene Warming and the American animal ecologist Professor Victor Shelford. These were followed, during the 1930s and 1940s, by important works by Dr. F. Clements, Professor V. Shelford, Professor J. Weaver, Professor J. R. Carpenter, and Professor L. Dice in America, and by Professor A. Tansley in Britain. This type of work is represented, at the present time, by studies comparing the structure and productivity of biomes rather than merely their description. In the early 1970s, a large international cooperative research program called the International Biological Program (I.B.P.) was organized. Linking the work of scientists from many countries, it aimed to gather a large body of comparable information on the structure and productivity of all the major biomes in various parts of the world. Although this program ended in 1974, the results of this immense amount of work will not be fully analyzed for a number of years to come.

The descriptions of the major biomes given in the following paragraphs are therefore not the work of one man but a synthesis of the works of many people over a considerable period of time.

The *tundra* (a word that comes from the Russian for "marshy plain") is the major vegetation type of the immense Arctic region. It occurs wherever the ground is permanently frozen up to within three feet or less of the surface (forming the *permafrost*.) This has the obvious consequence that there are no trees in the tundra, because no roots can properly grow in the permafrost. Furthermore, because of the solid ice just below the surface, water cannot easily drain away, and so, although very little rain falls, the ground is usually waterlogged during the short summer (and frozen hard, of course, in winter). Naturally, then, the tundra is not a hospitable habitat. Its simple communities contain few plant species. Of these, grasses and sedges predominate, especially in very marshy areas; the drier areas usually have some dwarf shrubs, such as heathers and bilberries. During the brief summers, Arctic-alpine flowers add some charm to the scene, and great patches of reindeer moss and other lichens thrive.

The *taiga*, a very extensive biome, extends across large areas of Canada and northern Eurasia. Its dominant plants are coniferous spruces, firs, and pines, but such deciduous trees as birches, poplars, and aspens can survive in the wetter southern fringes. The climate of the taiga is not unlike that of the tundra, with average monthly temperatures remaining below freezing for six months at a time. There is even a permafrost in the more northerly parts, although the ice is far enough underground to allow the growth of trees. The main climatic difference from the tundra is that the summers tend to be rather warmer and wetter.

In the temperate region to the south of the taiga, two main biomes exist, the temperate deciduous forests and the steppe grasslands. These regions are characterized by higher average temperatures and less seasonality, although the seasons are still very distinct. They can be differentiated by the amounts of rainfall. The deciduous forests occur in the wetter regions, which tend to be near the edges of continents, where moist winds come off the oceans carrying rain. The grasslands occupy drier regions in the more central parts of the continents. It is worth noting that these are the two biomes that have been most affected by mankind, particularly through agriculture.

Still in the temperate zone, but where the climate is "Mediterranean"—in California, for instance, and the countries bordering the Mediterranean—occurs the relatively small biome known as the *chaparral*. This is characterized by such drought-resistant shrubs as the manzanita and chamise. Although there are significant amounts of winter rainfall in the chaparral, the long summer drought precludes the growth of trees except along watercourses.

Moving further southward, we come upon the large desert biome. Here the rainfall is very low and irregular in both winter and summer. The only plants that can live under these conditions are drought-resistant perennials—cacti, and shrubs such as sage brush and creosote bush—and ephemerals, which spring up after the rare showers and flower, fruit, and die within a week or two. Closer to the equator, the arid desert merges into savanna grassland, where temperatures are relatively high all year long; although there are significant amounts of rainfall, they occur largely in the summer. This means that trees, with their continuously high water demand, cannot survive except in places where the water table comes close to the surface. The typical savanna scene is one of a sea of tall grass, punctuated with an occasional clump of acacias or baobab trees. Finally, we arrive at the three tropical-forest biomes—scrub forest, deciduous forest, and rain forest, in that order. All these regions have high, constant heat and heavy rains, providing an ideal climate for vegetation. It is here that we find the greatest diversity of trees, other plants, and animals.

The different types of climatic and vegetational regions are generally known; after all, any casual traveler can see and record them. But it is to C. Hart Merriam and the several other workers who followed him that we owe a definitive classification that permits the systematic study of the earth's biomes. (The word *biome* is a useful concept embodying the ideas of the interaction of climate, plants, and animals to produce a distinctive type of ecosystem, and was coined by Clements and Shelford in their important textbook *Bio-ecology* published in 1935.)

There are, however, many other ways of characterizing vegetation. One of these is to classify them according to their pattern of growth. The first important steps toward devising a system of this kind were taken by a Danish botanist, Christen Raunkaier, in 1903. He distinguished between five different types of plant growth (or "life forms," as he called them), based on the relative position of the resting bud. This is the bud that lies dormant during the plant's resting season, and eventually gives rise to the new season's growth.

Some plants have no resting bud at all, because they are either annuals or ephemerals, and therefore survive from year to year by seed. Raunkaier named them the *therophytes* (from the Greek words for "summer" and "plant"), to describe plants that are visible only in the summer. To a second life form, comprising plants with rhizomes, bulbs, corms, or tubers, in which the bud is buried beneath the surface, he gave the name *cryptophytes* ("hidden plants"). Those that have their buds right at the soil surface, such as tussock- and rosette-forming plants including grasses and daisies, are *hemicryptophytes* ("half-hidden plants"). The fourth category consists of the *chamaephytes* ("low plants"), whose buds are close to, but above, the ground surface, as in various creeping plants. The last of the five groups is the *phanerophytes*, in which the bud is well above the ground ("visible plants"), as in shrubs and trees.

Clearly, the position of the resting bud is related to the ability of the plant to withstand difficult conditions. Thus, plants with well-protected buds—the cryptophytes and hemicryptophytes—are more characteristic of the harsh tundra than of tropical forests, where the highly exposed phanerophytes abound. The therophytes, as we have already seen, are particularly abundant in the desert biome, where conditions are perhaps harshest of all. Their drought-resistant seeds give them a much better chance of surviving long periods without rain-

fall than a resting bud would have, no matter how well protected it might be (seeds, for example, can have thicker coats and hence reduce water loss).

An obvious limitation of Raunkaier's classification is that its concentration on the resting bud gives us no taxonomic picture of plants during the growing season. This was provided by two later ecologists, Pierre Dansereau, a French Canadian who did much of his work in America, and Charles Elton, for whom the Eltonian pyramid of numbers is named. Dansereau's rather complicated system divides growing vegetation into six categories. In the first place, he considered the plant's overall growth habit—does it grow as a tree, as a shrub, or as a herb? Secondly, the plant is classified according to its size; is it tall, medium, or low? The third category is based on a simple distinction: is the plant deciduous or evergreen? The fourth and fifth categories are leaf shape (such as needle, broad, or grasslike, simple or compound) and leaf texture (such as filmy, membranous, or succulent). Lastly, the vegetation is classified by the way that it covers the ground, whether sparsely, in tufts, or continuously. Almost any plant can be readily fitted into this all-embracing scheme of classification, which offers a highly useful set of clear criteria for describing the physical structure of virtually every community of plants.

Charles Elton's system of classification has one very important difference from all those considered so far. Although it deals with the structure of plant communities, it does so from the standpoint of the kinds of habitat that they provide for *animals*. Once again we return to a scene that figured largely in an earlier chapter of this volume—Wytham Wood, near Oxford in England. It was in the course of making a painstaking survey of the plants and animals in this wood that Elton saw the need for a general classification of all possible terrestrial and aquatic habitats of animals. His system is based on the fact that animals and plants go together— that once you have identified and classified the vegetation of a given area, you can do a good job of predicting the kinds of animals that are

Raunkaier's classification of vegetation is based on the position of the resting bud in relation to the soil surface. The histograms (right) show that the relative abundance of the different life forms is related to the harshness of the environment.

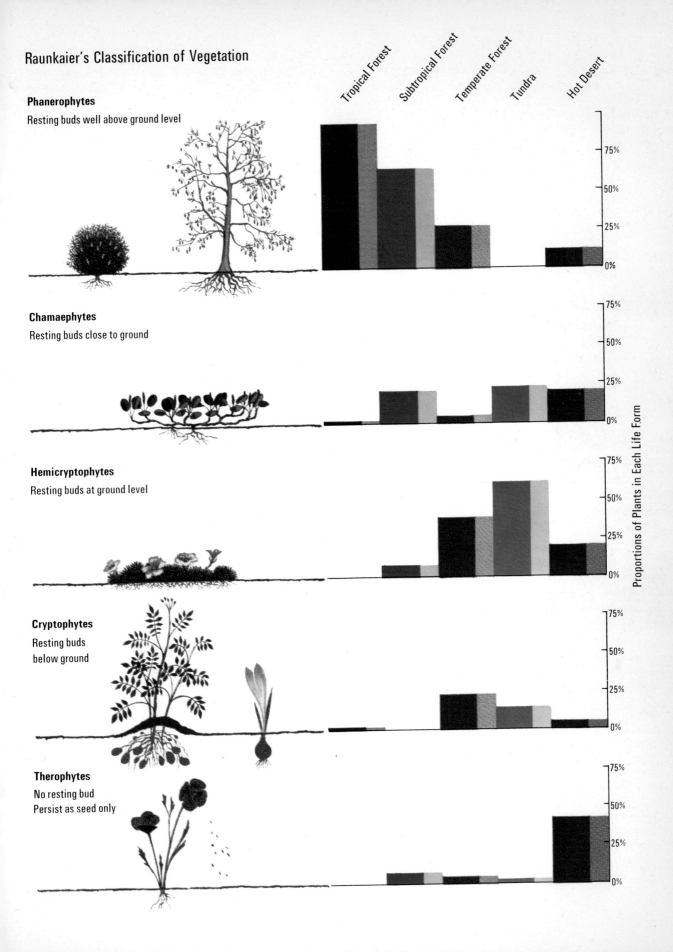

Raunkaier's Classification of Vegetation

Phanerophytes
Resting buds well above ground level

Chamaephytes
Resting buds close to ground

Hemicryptophytes
Resting buds at ground level

Cryptophytes
Resting buds
below ground

Therophytes
No resting bud
Persist as seed only

Tropical Forest

Subtropical Forest

Temperate Forest

Tundra

Hot Desert

Proportions of Plants in Each Life Form

75%
50%
25%
0%

75%
50%
25%
0%

75%
50%
25%
0%

75%
50%
25%
0%

75%
50%
25%
0%

likely to be found there. In the first place, Elton distinguishes among such broad categories as terrestrial, subterranean, aquatic, and semi-aquatic communities. He then divides each of these into a number of appropriate habitats, which he subdivides further as necessary.

Let us take, for example, the terrestrial system. Elton's system breaks this down into four categories—woodland, scrub, field, and open ground—and each of these is again broken down in accordance with its vertical structure. Thus, to take just one category, woodland can be divided into seven distinct zones. To begin with, there is a subsoil zone (below nine inches) followed above by a topsoil (from nine inches deep to the surface), where numerous microorganisms, insects, mites, and earthworms live. Above ground level there is a ground zone (up to six inches), a field layer (six inches to six feet), a low canopy (six to 15 feet), and a high canopy (15 feet to the treetops). The seventh and last category includes the air above the tree, which is a habitat not only for high-flying birds but for many insects and various organisms that comprise the aerial plankton.

The great merit of Elton's classification is that it provides a convenient shorthand for pinpointing the location of animal specimens that we collect in any kind of biome. Thus, winter moth females may be caught anywhere between the ground and the high canopy, whereas the caterpillars will occur only in the low and high canopies or in the air above when floating on their silken threads. Winter moth pupae, on the other hand, are found only in the topsoil zone. In effect, the vertical system is designed to group together all the animals that are likely to come into direct contact with one another. This is a necessary preliminary step toward determining the structure of food webs.

The classification of habitats, then, can be seen as a significant part of the imposing body of scientific knowledge about the natural world that ecologists have been gathering over the past century. This body of knowledge is not merely impressive but also performs an extremely important function. The ecologists whose achievements we have described in this book have tackled a very great variety of problems, ranging from detailed population studies of a single species to descriptions of whole communities. Their approaches have often been very different, but their ultimate aim is always the same: to understand the relationship of animals and plants to one another and to their physical habitat. As the understanding of such relationships improves, we become better able to manage our natural resources without causing irreparable damage to the communities around us. And so ecology provides the basic knowledge that is necessary for the battle to preserve the richness and diversity of natural systems for the good of future generations.

Elton's vertical classification of habitats, shown here for an oak wood, provides a useful description of the characteristic habitats of animals, such as the feeding levels of birds.

Habitat Divisions in an Oak Woodland
Based on Elton's Classification

1 Swallow	9 Robin
2 House martin	10 Treecreeper
3 Pied flycatcher	11 Garden warbler
4 Spotted flycatcher	12 Nightingale
5 Wood warbler	13 Wren
6 Chiffchaff	14 Hedge sparrow
7 Redstart	15 Blackbird
8 Willow warbler	

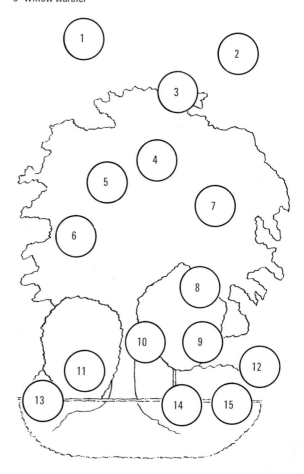

Air Above

High Canopy
(15 feet to the
treetops)

Low Canopy
(6 to 15 feet)

Field Layer
(6 inches to 6 feet)

Ground Layer
(0 to 6 inches)

Topsoil Layer
(0 to 9 inches deep)

Subsoil Layer
(Below 9 inches)

63

Index   Page numbers in *italics* refer to illustrations or captions to illustrations

Picture Credits

Key to position of picture on page: (B) bottom, (C) center, (L) left, (R) right, (T) top; hence (BR) bottom right, (CL) center left, etc.

10–1   David Nockels and Alan Hollingbery © Aldus Books
13   Photo Roger Hyde © Aldus Books
16   David Nockels © Aldus Books, after G. C. Varley, G. R. Gradwell, and M. P. Hassell, *Insect Population Ecology, an Analytical Approach,* Blackwell Scientific Publications Ltd., Oxford, 1973
17(R)   Photo Professor G. C. Varley
18–9   Photos Roger Hyde © Aldus Books
20–1   David Nockels © Aldus Books
22   Alan Hollingbery © Aldus Books
24–5   Biophoto Associates
26–7   Alan Hollingbery © Aldus Books, after G. F. Gause, *The Struggle for Existence,* 1934, The Williams & Wilkins Company, Baltimore
30–1   Alan Hollingbery and David Nockels, after C. B. Huffaker, *Hilgardia,* 1958, Vol. 27, pp. 243–383, © Aldus Books
33   David Nockels © Aldus Books

34   Dick Clarke/Seaphot
35(R)   Heather Angel
36   British Crown Copyright. Reproduced by courtesy of the Controller of Her Britannic Majesty's Stationery Office
37   Seaphot
37(R)   Alan Hollingbery © Aldus Books, after Harold Barnes, *Oceanography and Marine Biology,* George Allen & Unwin Ltd., London, and Hafner Publishing Company Inc., New York
39   David Nockels © Aldus Books, after E. P. Odum, *Ecology,* Holt, Rinehart & Winston, New York
42–3   David Nockels © Aldus Books, after information contained in J. M. Teal, *Ecological Monographs,* Vol. 27, pp. 283–302, 1957, Duke University Press, Durham, North Carolina
44(B)   Heather Angel
45   Picturepoint, London

46   David Nockels © Aldus Books, after C. M. Woodwell, "Toxic Substances and Ecological Cycles," March, 1967. © by Scientific American, Inc. All rights reserved
47   K. Fink/Ardea London
49   Heather Angel
52–3   David Nockels and Alan Hollingbery © Aldus Books, after information contained in B. Maguire, *Ecological Monographs,* Vol. 33, pp. 161–85, 1963, and D. S. Simberloff and E. O. Wilson, *Ecology,* Vol. 51 pp. 934–7, 1970, Duke University Press Durham, North Carolina
54   David Nockels and Alan Hollingbery © Aldus Books, after graphs in Robert H. MacArthur and Edward O. Wilson, *The Theory of Island Biogeography,* Princeton University Press
55(T)   Alan Hollingbery © Aldus Books
57   Photo Roger Hyde © Aldus Books
61–3   David Nockels © Aldus Books